Durable Concrete: Build Namibia

Shah

First Printing, 2024

Contents

1. Introduction

1.1. Background to the research

The Namibian construction sector is an essential portion of the country's economy with concrete construction being key to it. It is thus vital that the quality of concrete design and construction should be appropriately managed. Namibia faces different challenges related to concrete durability. In the coastal areas (including Walvis Bay and Swakopmund) rebar corrosion is a common and visible problem on structures, due to the prevailing environmental conditions. Similarly, bridges and road culverts, in the northern parts of the country have shown corrosion damage. Most of the inland of Namibia is a desert area where extreme weather conditions provide challenges to concrete casting and ultimately durability.

Currently, Namibia does not have local concrete durability design and construction specifications but follows international design codes and specification (South African and British standards) which are prescriptive approaches. In this approach, the achievement of minimum strength (with limiting values of material and construction methods) is believed to produce durable concrete. Studies undertaken internationally have shown that the prescriptive approach has shortfalls (Santhanam, 2013). Though environmental exposure is considered, the approach relies on providing limiting values on concrete material and mix proportions to produce durable concrete. In the process, the approach ignores other factors such as binder/aggregate types and on-site construction practice which influence concrete durability in any environment (Beushausen, 2014). Concrete durability has been shown to improve considerably with the design and construction specifications that expressly point to the performance required of the concrete. This involves having measurable concrete qualities and testing capabilities on these qualities (Alexander, Beushausen, Torrent, 2008).

Concrete deterioration due to reinforcement corrosion has been shown to be the principal cause of early concrete maintenance needs globally, accounting for as much as 5% of many countries' GDP (Alexander, 2016). To improve on concrete durability, some countries have developed concrete deterioration models that are used to predict how long concrete can perform in a defined environment. Countries in Europe were the first to develop mathematical concrete deterioration models, under the DuraCrete project, with which engineers could assess concrete performance in its life time (Siemens and Edvardsen, 1999). Different deterioration models have been developed in a number of countries; Canada, USA, India, Australia and (within the region) South Africa. The models developed measure transport-related properties of a concrete that show how long concrete would deteriorate either due to carbon-induced or chlorine-induced corrosion.

In South Africa, mathematical models generate durability indices to show concrete resistance to ingress of chloride and carbonation (Beushausen, 2014). Three different durability index test regimes have been developed. These test regimes have been so successful that they have been accepted as part of the South African standards for testing concrete works. The regimes have been used in assisting with improving concrete mix design, quality control, conformance monitoring and as-built concrete monitoring. Quality control is the entire process of ensuring that concrete produced is of acceptable quality standard. Conformance monitoring being a procedure which shows that produced concrete meets the set quality standards whereas, as-built monitoring relates to verifying the quality performance of built concrete structures/elements.

Various concrete structures have been built across Namibia as part of public infrastructure and the durability of these structures is vital to the economy. However, there currently lacks database (national or local) on the state of infrastructure that can show what durability problems are common where. This research was undertaken in part to contribute to this knowledge on common defects in concrete structures across the nation based on sampled structures. The research will also contribute to the body of knowledge on standards and specification being used in Namibia's concrete construction including challenges and benefits of using them.

The research was undertaken through literature review, condition assessment and interviews. Literature review was split into three parts;

a) Review of the regulatory framework to understand effects of any proposed changes on the existing framework.
b) Review of local technical documents on concreting practices.
c) Document recent development in performance-based design and construction specification for concrete works in South Africa

Interviews were conducted with practitioners to get an understanding of industry knowledge on durability issues. Designer/specifiers, producers, quality controllers and infrastructure owners were interviewed (Appendix A).

Concrete condition assessment was undertaken on structures in four towns in Namibia. Selection of the structures was driven by access to undertake an inspection, access of managing office amongst others. It was also decided to look at types of structures which could be found in multiple areas and thus be comparable. The assessments looked at the concrete deterioration and, based on the literature review, identified factors that had led to the deterioration.

1.2. Research problem statement

The availability of information on the defects of structures (for specific areas) is important in design and management of concrete structures as it provides a reference for need of better practice. Where defects cannot be presented it is easy for developers, engineers and managers to believe everything is alright. Currently, there is minimal academic documentation on the condition of concrete structures in Namibia. It is believed that the results of this research can in future be used as a platform for further research in durability of concrete structures and the standards that govern/manage production of durable concrete structures. To this end the research expects to

a) Document common concrete defects in the project areas
b) Identify and analyse commonly used concrete design and construction specification with specific look at how these aid (or do not aid) in production of durable concrete
c) Identify improved international practices in durability concrete design and concrete structure management that would add value to concrete practice in Namibia

1.3. Research limitation

In undertaking the research the following challenges were encountered which limited the data collected

a) A number of officers approached were not forthcoming with information that would have helped in answering some question. Information like the maintenance undertaken on some bridges and water tanks was not supplied. This left the author to not understand why these were undertaken
b) Non-destructive testing of structures was not undertaken due to the financial implication of undertaking them. As most laboratories do not normally undertaken these measurements requesting for the test was at a premium which the author could not afford being a self-funded research.
c) Sites assessed were limited to areas close to the author's base station because of limited financial capability.

2. Namibia the country landscape and the concreting environment

2.1. Introduction

Namibia is an extensively vast country with a total land area of 825,000 sq. Km (Christelis and Wilhelm, 2011). It is a country with diverse topography and climate. It lies on the south-western edge of Africa with the Atlantic Ocean on the western border and South Africa, Zambia, Botswana and Angola on the other borders (Fig. 2-1). The Atlantic Ocean influences the weather and climatic patterns prevalent in the country and the countries around influence the concrete construction practices (either through material imports or technical knowledge). South African technical expertise and materials are used in most construction works meanwhile Zambia, Angola and influence mostly materials that are used in concrete.

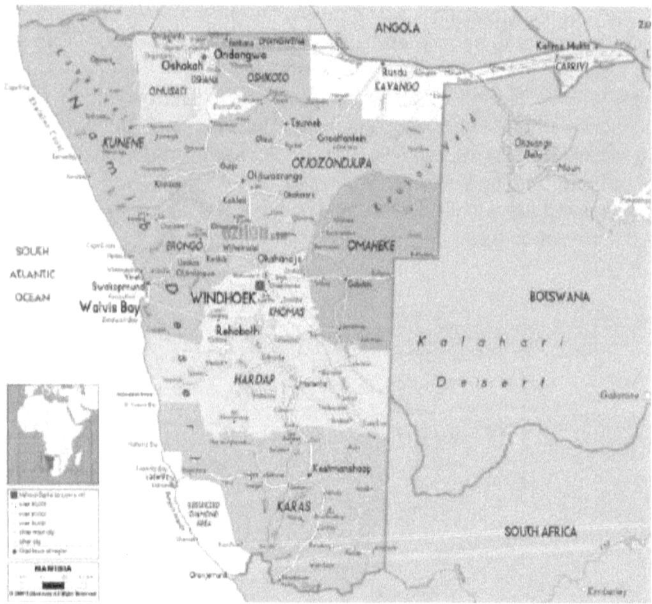

Fig. 2-1: Administrative map of Namibia (Source: www.ezilon.com, 2017)

The varied geographical conditions in the country present challenges to the concreting process and durability of built concrete structures. Durability has been defined by Alexander, Ballim, Beushausen (2009) as 'the ability of a structure or component to withstand the design environment over the design life, without undue loss of serviceability or need for major repair'. Durability research has enabled the global concrete industry to better understand the effects of the environmental conditions on built concrete structures (Siemens and Edvardsen, 1999). It is thus vital to understand the Namibian concreting environment (i.e. weather/climate, concrete design and construction practices and legal framework) to be in a position to advise on durability issues in Namibia.

The political administration of the country is managed by a central administration which has devolved powers to regional and urban councils. These councils have powers to regulate construction works within their boundaries (the regional administrations are shown in Fig. 2-1). The central government manages both government departments and state-owned institutions that either are involved in concrete construction or regulate construction works. Political administration provides legislation which influence works execution/management, which is why it is essential to understand the structure and powers.

2.2. Weather and Climate

The weather and climate in Namibia are heavily influenced by the cold Benguela current flowing in the Atlantic Ocean on the western border of the country (Byers, 1997). This edge of the country is part of the Namib Desert whereas the Kalahari Desert covers the south and parts of the south-east. Namibia is mostly an arid country with more than two-thirds of land receiving less than 400 mm of rain annually. The air is generally dry with most portions of the country having relative humidity less than 50%. The mean temperature across the country varies; the south-west with lowest mean temperature of 11.2^0 C and the north having the highest mean temperature of 23^0 C (Kudzai et al 2016). These weather variations are shown in Fig. 2-2.

To the west, Namibia borders the Atlantic Ocean, which Aquafact (2012) documented to experience large swells with waves up to 2.5 m. The water has a salinity level of up to 35.5%, which is within the average salinity levels in the oceans of the world (Aquafact, 2012). This area receives very little rain; however, the cold air coming from the Atlantic coast generates a fog which is responsible for the precipitation measured. This fog influence extends to as much as 100Km inland (Lancaster, 2002). These conditions and the high relative humidity imply concrete structures built are in continuous exposure to chloride-saturated water that is rarely removed by any naturally occurring rains. The effect has made the coast a very corrosive environment for steel. Callaghan (1991) undertook a 20-year study of corrosion due to atmospheric conditions in southern Africa from which an atmospheric corrosion map was produced (Fig 2-3) which highlights the corrosion risk in parts of Namibia.

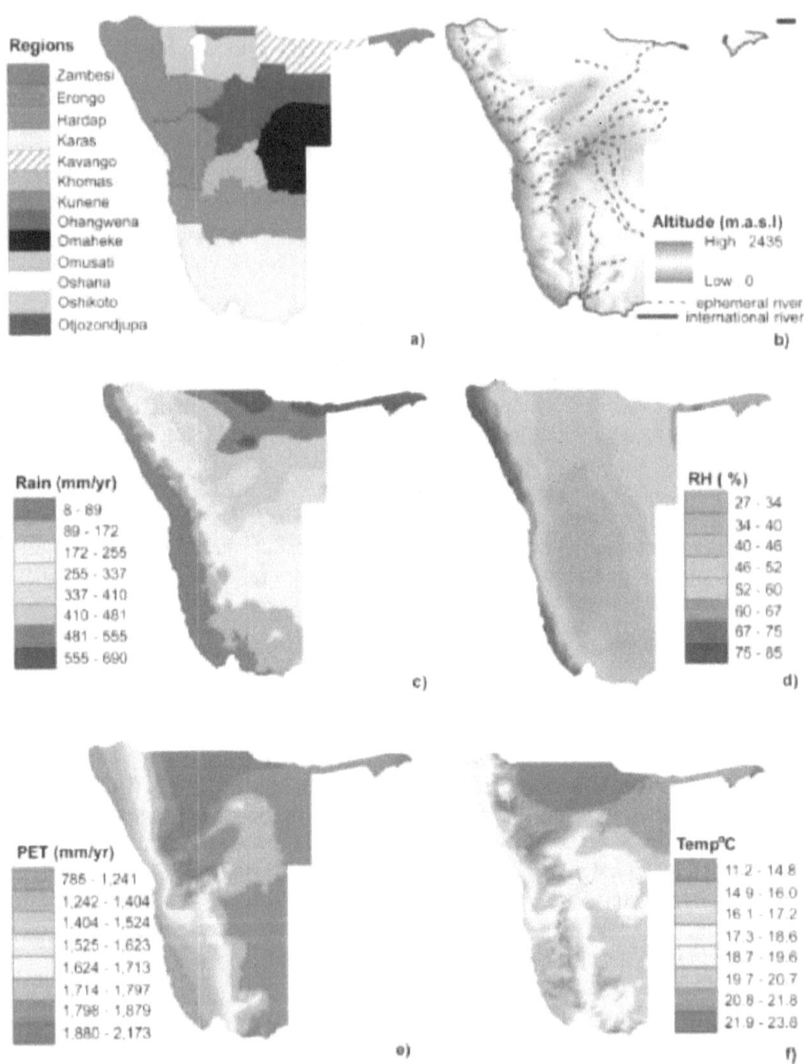

Fig. 2-2: Namibia geographical characteristics (Source: Kudzai *et al*, 2016)

LEGEND

Code	Description	Map identif.	Type of corrosion	Mild steel* corrosion rate µm/yr	Galvanised steel sheet** life in years'
A	Intertidal to 5 km inland		Severe marine	100 - 300	Up to 3
B	Desert marine (Mists)		Severe marine	80 - 100	0.5 - 2
C	Temperate marine		Marine	30 - 50	3 - 7
D	Sub-tropical marine		Medium to severe marine	50 - 80	3 - 5
E	Desert inland dry		Desert	< 5	> 30
F	Inland		Rural	10 - 20	> 20
G	Inland urban		Inland industrial[1]	15 - 40	5 - 15
H	Urban coastal		Marine industrial[1]	50 - 150	1 - 3
I	Inland arid		Semi desert	5 - 10	> 30

FIG. 2-3: Atmospheric corrosion map for southern Africa (Callaghan, 1991)

2.3. Topography and geology

The central part of Namibia is the highest portion of the country rising to 2435 m above sea level. The country has a conical profile with the land falling from the central region to all the border regions. As Fig. 2-2b shows, the western border has the lowest elevation sitting mostly at sea level.

The geology of Namibia has a distinct age difference between rocks in the west to those in the east, with the east having the youngest rock formation (Goudie and Viles, 2015). The primary influence of this geology has been on surface and groundwater profiles of the country, as shown in Fig. 2-2b and Fig. 2-4. For the surface water, Fig. 2-4 shows that virtually all rivers in the country only flow for a short period and large areas of the country have great potential for groundwater. It can also be seen that the presence of saline groundwater is widespread across the country. The presence of this saline water is a significant risk of corrosion of reinforcement embedded in concrete.

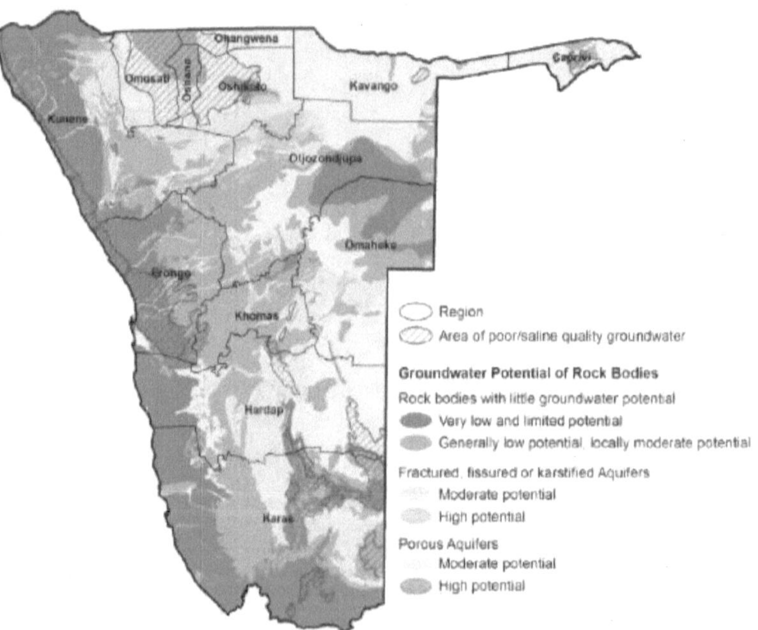

Fig. 2-4 : Namibia Hydrogeological profile (Source - Christelis *et al*, 2006)

The geology of the country has within it substantial mineral resources. For the concrete industry, several large limestone deposits exist in the country which are utilized in cement production. For a long time, Namibia imported cement for the construction industry, but from 2017 domestic cement production has been enough for domestic cement demand (Namibia economist, 2016). Local cement production has provided the construction sector an opportunity to explore cement which could best be suitable for the localised environment. Ohorongo Cement mines limestone, shale and marl to produce CEM I and CEM II cements (Ohorongo, 2019) whereas Cheetah cement produces only CEM II cement (with imported clinker). CEM I cement has more than or equal to 95% Portland cement clinker in its composition whereas CEM II cement has additional materials besides clinker (in the Namibian case additional materials ranging from 6% to 35% are used).

2.4. Regulatory framework

Construction work in Namibia is governed by several laws and regulations that impose minimum standards to be met by every development. Principal to these are the town planning regulations (Town Planning Ordinance of 1954, the Townships & Division of Lands Ordinance 11 of 1963 and the Local Authorities Act 23 of 1992). These laws dictate which authority will impose regulations that any new development has to comply with (Local Authorities Act 1992). In recent years, there have been ongoing attempts to update the national building regulations and ensure uniformity in national regulation of construction works (Namibia economist, 2015).

Engineering practice in Namibia is regulated by the Engineering Council of Namibia in line with the Engineering Professionals Act 1986 as amended in 1991. The council, amongst other things, ensures adhesion to minimum accepted engineering practice by individuals practising as engineers. It also liaises with various engineering professional bodies interested in specific engineering disciplines. These professional bodies include the body representing consulting engineers, contractors, and various others. Any attempt to implement new construction specifications or design standards will require the collective acceptance (or acknowledgement) of all stakeholders.

2.5. Conclusion

Namibia is a country with varied terrain, topography and climatic conditions. The climatic conditions are very extreme and present differing challenges to any concrete works and structure built in it. There are multiple levels of legislation which influence infrastructure works. However, most works are controlled by state owned enterprises which have powers to manage quality control of works. Coordinating changes in engineering practice should theoretically be easy due to to the fact that a single body manages and liases with all engineering practicing bodies.

3. Concrete durability, deterioration and testing

3.1. Introduction

Concrete is a composite material that deteriorates with time. The environment that concrete structures are built in has a direct influence on its durability. This influence occurs both at individual constituent material and composite concrete level. Production (and management) of durable concrete structures thus requires an understanding of general concrete characteristics, concrete defects, deterioration mechanisms and tests that can be performed.

Factors influencing durability of concrete structures can be grouped into those factors related to the concrete and the environment of exposure. For the concrete, the factors can be directly linked to the concrete structure makeup (intrinsic factors) or the process of creating the concrete structure (extrinsic factors). Environmental factors relate to the aggressiveness of the environment, which can be either physical or chemical factors. This is best shown in Fig. 3-1.

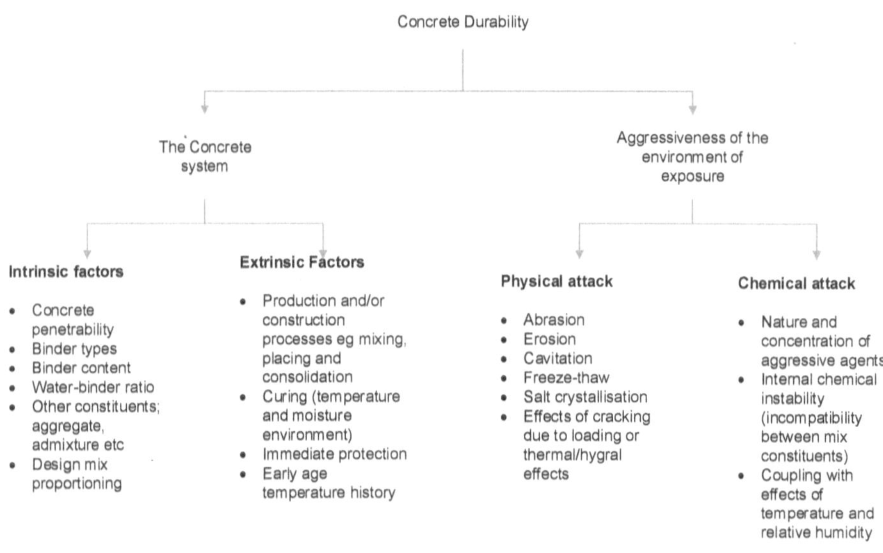

Fig. 3-1: Concrete durability factors (Alexander, Ballim, Beushausen, 2009)

3.2. Concrete properties and durability

Virtually, all processes that lead to concrete deterioration are related to fluid ingress and movement into and within the concrete. The manner of movement of fluids (in and out of concrete) are termed transportation mechanisms.

Penetrability is a critical characteristic in concrete fluid movement. Alexander, Ballim, Beushausen (2009) defined penetrability as the degree to which fluids and ions can ingress into and move through concrete. In this definition, it is appreciated that different substances (liquids, gases and ions) move under different forces. The movement can be due to pressure difference, concentration gradient or capillary action.

Permeability in concrete is a measure of the movement of liquid through saturated concrete pores under the influence of external pressure (Alexander, Ballim, Beushausen, 2009). The concrete pores, in this case, would be saturated with water. The water movement is a function of porosity and interconnectedness of the pores (Li, 2006). Concrete can have a higher porosity but a lower permeability simply because the pores are not interconnected. The pores are found at the interfacial transition zone (ITZ) and also between cement particles. As hydration progresses, the C-H-S matrix closes out the available pores. In Fig 3-2, the relationship between porosity and permeability is exhibited. In the figure, both concrete have a similar porosity, but concrete (a) has a higher permeability than concrete (b) due to the interconnectivity of the pores.

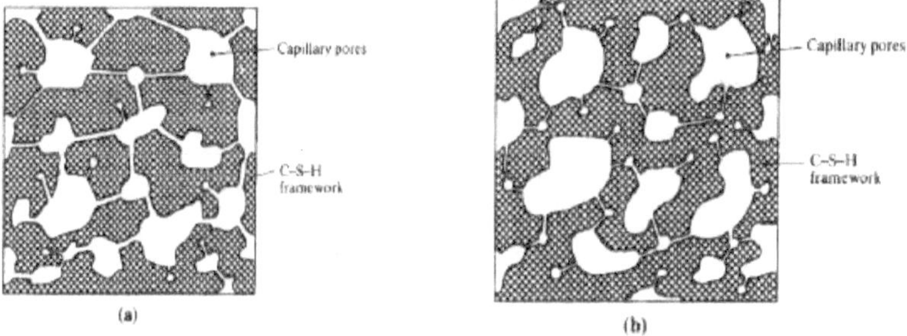

Fig. 3-2: Porosity and permeability in concrete (Source: Neville & Brooks, 2003)

Where there is a concentration difference between internal or external concrete environment (or even within the internal spaces), ions or moisture or gases will move due to a concentration gradient. This movement process is called diffusion. Different fluids and ions will have different diffusion coefficients, and even for the same element, its diffusion is influenced by temperature, its interaction with concreting products and moisture content in the concrete. Diffusion will occur in an environment that can be partially or fully saturated. Most salts move by this transport mode, and thus it is an essential mechanism in the durability of concrete.

3.3. Physical attack

Physical attack involves direct action on the concrete structure. The action can be a result of external environmental action or internal chemical reaction that produces a force on the concrete. Abrasion, erosion and cavitation are all due to action by the external environment on the concrete. In all these cases, the ability of concrete to resist being stripped off (abrasive resistance) is a function of the strength of the concrete (IS 536). Alexander, Ballim, Beushausen (2009) have shown that a relationship exists between water/binder ratio and increase in abrasion resistance.

In the marine structures concrete elements that are in a state of wetting and drying due to tidal action, salt crystallization is common (Alexander, Ballim, Beushausen, 2009). When the wetting period occurs, sea water (which is essentially a salt solution) is absorbed into the concrete through its surface. At low tide and with increased temperature water evaporates, from the element, leaving the salt in the concrete. When the next cycle of wetting occurs, more salty water gets absorbed. Thus more salt is available to crystallise on evaporation of the water. As this cycle continues, the crystallised salt can cause expansion to occur, leading to cracking in the concrete surface. Concrete with low permeability will ably avoid this problem. Concrete structures along Namibia's coastline are under real threat of the salt crystallization problem.

3.4. Chemical attack

Chemical attack involves aggressive agents reacting with elements in concrete composite structure affecting the concrete's performance. Alexander, Ballim, Beushausen (2009) presented a schematic of concrete chemical attack summary. In the schematic, three different mechanisms were identified (Fig. 3-3). Meanwhile, Bijen (2003), identified four different attack mechanisms. The difference between the two is that the latter includes alkali-silica reaction as a separate attack mechanism.

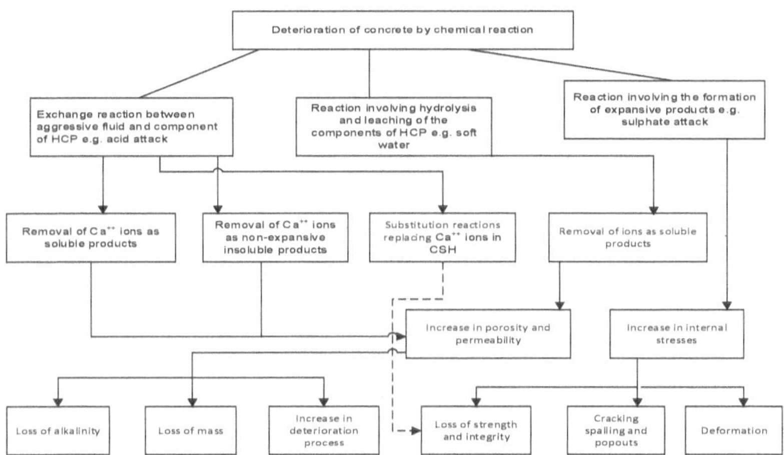

Fig. 3-3: Summary concrete chemical attack schematic (Source: Ballim *et al,* 2009)

As Fig. 3-3 shows, some of the major effects of chemical attack are increase in porosity, permeability and cracking of the hardened concrete. Due to these effects, the risk of reinforcement embedded in the concrete corroding and thus drastically reduce the service life of the structure built. Almaral-Sanchez et al (2011), in a study on sulphate attack and reinforcement corrosion in concrete using recycled material, showed that changes in cementitious material increased the resistance to sulphate attack and reinforcement corrosion. For designers and construction engineers, an appreciation of prevalent aggressive chemical environment (for an area) can enable them better mitigate against chemical attack for planned structures.

3.5. Mechanical attack

Mechanical attack is where a load is imposed on concrete that is greater than its load-carrying capacity. This can be due to overloaded or impact on a concrete element. The extent of damage to the concrete is dependent on the duration of loading and nature of the load. Cyclic loads have a tendency to cause much more extensive damage than static loads.

3.6. Alkali aggregate reaction

Two forms of alkali-aggregate reaction have been identified in concrete; alkali-silica reaction (ASR) and alkali-carbonate reaction (ACR). In either case, the cement paste reacts with silica or carbonate compounds naturally occurring in the aggregates leading to expansive cracking of the concrete. Of the two, ASR is better understood (Li, 2009).

ASR is a chemical reaction that occurs in concrete with aggregate having reactive silica. When used in concrete, the silica reacts with alkalis in (that comes from the cement and other cementious material used). The compound formed dissolves in pore solution to form a silica gel which attracts water from around it (Bijen, 2003). The gel expands in the presence of water. The more silica there is, the higher the 'pull force' for the water and the bigger the expansion. This leads to cracking once the expansive force exceeds the concrete tensile capacity. Besides the unsightly cracked surface, the cracked concrete makes the reinforcement within more susceptible to corrosion due to ingress of water and air.

3.7. Reinforcement Corrosion

Reinforcement corrosion is not really a deterioration of concrete, rather the reinforcement in the concrete. However, the result of the reinforcement corroding leads to the cracking of concrete within which the reinforcement is encased. Reinforcement corrosion is an electrochemical process which occurs when reinforcement bars react with oxygen and water leading to degradation of both the reinforcement and the concrete. Reinforcement corrosion is acknowledged to be the most significant cause of concrete deterioration (Newman and Choo, 2008). In general, when reinforcement corrodes; an anode and a cathode develops on the reinforcement bar with pore-water acting as the electrolyte, this enables the movement of ions between the anode and the cathode. This is shown in Fig. 3-4.

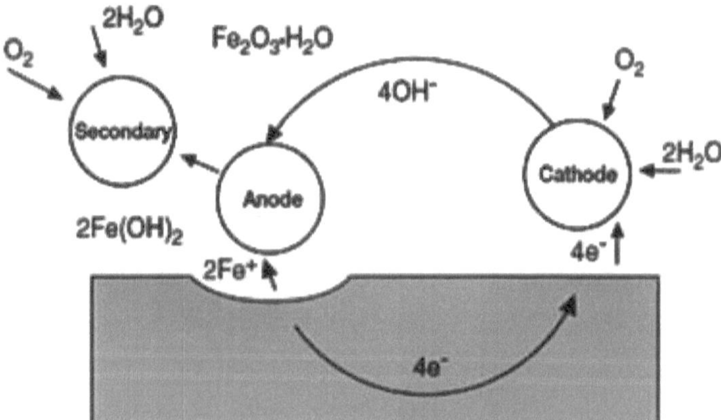

Fig. 3-4: A general steel corrosion schematic (Source: Portland Cement Association, 2002)

Steel reinforcement is made of a material whose surface does not have uniform physical and chemical characteristics due to either the manufacturing process (of the iron) or damage sustained prior to the steel being embedded in concrete. These non-uniformities lead to the creation of anodes and cathodes on the reinforcement surface before it is embedded in concrete. Once embedded in concrete and in the presence of oxygen and water under specific conditions, depassivation of steel occurs, which leads to the steel corroding. In concrete, there should not be any corrosion problem because the concrete environment is an alkaline environment with pH range 12.5-13.5 (Li, 2006). With the pH in this range, the reinforcement is passivated such that iron cannot dissolve into the pore solution, and it cannot react with oxygen. However, due to several factors, the pH in the concrete can be lowered, leading to corrosion of steel. The product of corrosion (i.e. the iron oxide) occupies a higher volume than the steel and thus as more corrosion products get produced the more volume is required. The expansion exerts pressure on concrete around the reinforcement. Once the tensile capacity of the concrete is exceeded, the concrete cracks leading to accelerated corrosion and more deterioration of the concrete.

Corrosion of steel goes through 3 phases: initiation, propagation and acceleration (Ciria C674). The initiation phase is from concrete construction to the time when corrosion starts and the steel is still passive, but there has been an ingress of aggressive agents. The initiation phase ends at depassivation of the reinforcement and propagation phase commences. In the propagation phase, there is significant corrosion of steel, but the steel has not reached ultimate state. Finally, in the acceleration phase corrosion is so severe that the corrosion products have expanded considerably, leading to extensive concrete cracking and eventually spalling. If the corrosion is not attended (to in this phase) the steel will rapidly reach ultimate limit state, which would reduce the structure carrying capacity (Ciria C674). The 3-phases of reinforcement corrosion are shown in Fig 3-5.

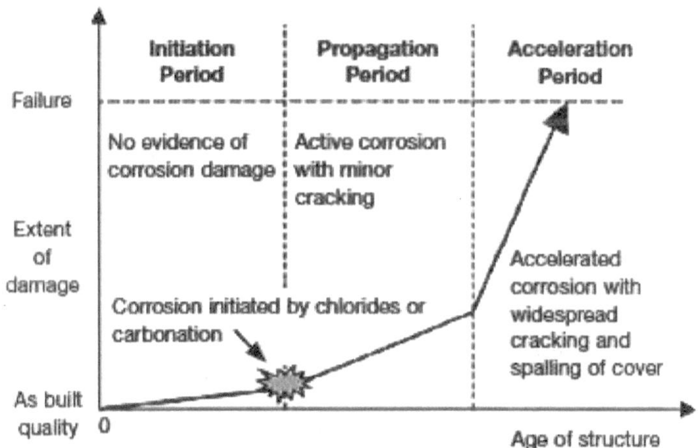

Fig. 3-5: The 3-phases corrosion damage model (Beushausen & Alexander, 2009)

There are two principal ways that the passive layer gets destroyed (depassivation) in concrete leading to reinforcement corrosion. These are carbonation and chloride contamination. There is also a form of corrosion that occurs in concrete structure carrying wastewater that is called biogenic (section 3.8).

3.7.1. Chloride contamination

Chlorides in concrete can either be introduced during mixing (from the water or aggregate contamination) or can come from the environment through diffusion into the hardened concrete. In concrete, chloride exists in two forms; as combined or free chloride ions. Combined chloride is joined to C_3A in the cement paste whereas the free chloride is suspended in the pore solution. The free chloride ions are the ones involved indirectly attacking the passive layer surrounding the reinforcement bars. For chloride ions to destroy the passive layer, the free chloride ions have to exceed the chloride threshold of the concrete. Chloride threshold value is a ratio of total chloride to binder content in cement expressed as a percentage. Internationally the point of measurement and the value are not yet agreed (Ueli et al, 2009). Whereas some define the chloride threshold value as measured at point when chloride reaches the reinforcement (depassivation) others define it as measured at point of noticeable deterioration of reinforced concrete element. Meanwhile the value itself ranges from 0.2% - 0.4%.

A characteristic of chloride-induced corrosion is that it directly attacks the passive layer around the reinforcement at a specific point while the rest of the passive layer remains intact. This leads to pitting corrosion (Newman and Choo, 2008) a main sign of chloride-induced corrosion. Fig. 3-6 is a schematic diagram of chloride-induced corrosion due to ingress of chloride ions in the form of common salt (NaCl). In the chemical reaction that leads to corrosion of the steel, chloride ions act as a mere catalyst in a multi-stage process. The salt dissolves in pore solution within the concrete matrix releasing sodium (Na^+) and chloride (Cl^-). Due to the electrochemical process (highlighted before) iron ions (Fe^{2+}) are released at the anode into the pore solution and react with Cl^-. The product of this reaction further reacts with water in the pore solution to produce hydrochloric acid, which reduces the pH of the pore solution (which normally is >12.5). The acid produced accelerates dissolution of iron ions and increases the amount of hydrochloric acid produced. The process continues as long as there are enough free chlorine ions and water within the matrix. The chemical reaction between chloride and iron is represented by equation 3-1.

$$Fe^{2+} + 2Cl^- \longrightarrow FeCl_2$$

$$FeCl_2 + 2H_2O \longrightarrow Fe(OH)_2 + 2HCl \quad \ldots\ldots\ 3\text{-}1$$

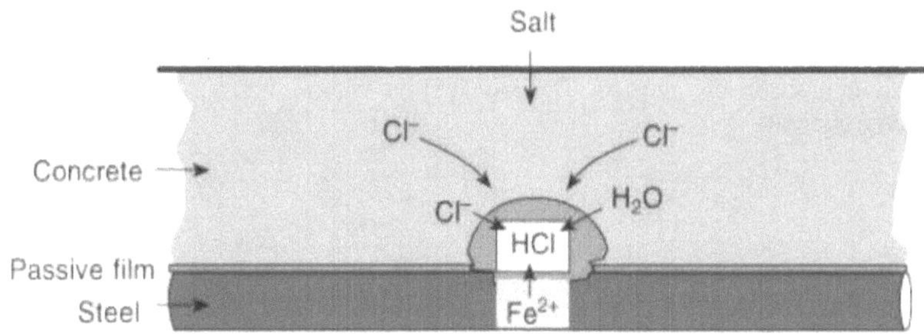

Fig. 3-6: Chloride-induced corrosion initiation schematic (Source Newman and Choo, 2008)

3.7.2. Carbonation

Carbonation is a process where atmospheric carbon dioxide diffuses and reacts with calcium hydroxide in concrete to produce calcium carbonate (Equation 3-2). The produced calcium carbonate has the effect of reducing the pH of pore water in concrete to below 10. This process occurs at the water-air interface in the concrete pores such that if the pores are saturated carbonation cannot occur (Bijen, 2003). As more carbon dioxide penetrates the concrete surface more of the calcium carbonate gets produced, in the process, this layer of calcium carbonate moves deeper into the concrete. The rate of production of the calcium carbonate gets reduced because the pores become more blocked with the calcium carbonate produced before (Perkins, 1997). This is why the front of carbonation moves slowly into the concrete. Once the front reaches the reinforcement level, the passive layer surrounding the reinforcement bars gets destroyed by the lower pH (of the carbonation front) leading to the development of anodic cells. In the presence of water and oxygen corrosion of the rebars occurs.

$$Ca(OH)_2 + CO_2 \rightarrow CaCO_3 + H_20 \; \ldots\ldots \; (3\text{-}2)$$

3.8. Biogenic concrete deterioration

Concrete is susceptible to deterioration due to actions of chemicals produced by microorganisms. A good example of such a deterioration is common in concrete carrying or holding wastewater. In this case when the wastewater becomes anaerobic microorganisms in it produce hydrogen sulphide (H_2S) and carbon dioxide (CO_2) which settle on concrete surface above the wastewater line. The carbon dioxide reacts with water to form carbonic acid (H_2CO_3). The carbonic acid reacts with concrete leading to carbonation but more damaging is that it reacts with the hydrogen sulphide to produce biogenic sulphuric acid. The sulphuric acid chemically attack the silicate hydrate (in the cement paste) and calcium hydroxide (in lime) to produce gypsum and ettringite. The formation of these two lead to loss of material creation of voids and loss of concrete strength due to expansive nature of these products, Fig 3-7 diagrammatically present the chemical process in a concrete pipe (Ng & Kwan, 2015). Haifen et al (2015) highlighted the pH sensitivity of the degradation especially as ettringite does not survive for long at pH lower than 11 (and yet in the anerobic conditions pH are usually below 11) then it decomposes into gypsum. This highlight the important role of gypsum in concrete degradation.

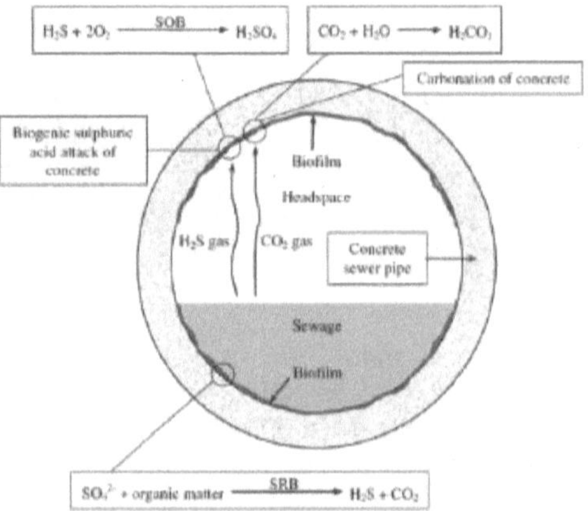

Fig. 3-7: Biogenic concrete deterioration in a concrete wastewater pipe (Source: Ng & Kwan, 2015)

Yarong et al (2020) and Yarong et al (2018) carried out investigation on the corrosion of reinforcement in sewer pipes and reinforced concrete structures. The two investigation confirmed that the corrosion of reinforcement bars, in sewers, is not the same electrochemical process as described in chapter in 3.8. The studies established that H_2S and chloride play a big role in the corrosion of reinforcement bars, with H_2S aiding in the liberation of oxygen from oxides and chloride assist in initiating the steel corrosion. Whereas oxygen played a role at initiation of corrosion the studies showed that over time sulphur and calcium play a more important role.

3.9. Concrete cover and durability

The ingress of aggressive agents into the concrete after casting is responsible for its deterioration and thus a reduction in durability. The ability of aggressive agents to penetrate concrete from the outside is a function of the quality of the concrete that resists the ingress. This is the concrete between the structure surface and just above the reinforcement normally called 'cover'. The importance of concrete cover can be seen in how design standards and specification emphasise on achieving a minimum thickness of this concrete. Mechanism discussed in 3.3 to 3.8 are generally responsible for destruction of the concrete cover. Once the concrete cover cracks there is an accelerated deterioration of the concrete. This is because the ingress of aggressive chemical agents is influenced by the crack properties more than the initial transport mechanism that led to the cracks (Alexander, Ballim, Beushausen, 2009). Thus for concrete durability concrete cover integrity is very important.

3.10. Concrete exposure conditions

Researchers long appreciated that the environmental effect on concrete durability varies with how aggressive an environment is. This is why in the design of reinforced concrete, virtually all design codes have provided guidance on dealing with aggressiveness of the environment the concrete will be built into. Previously the definition of these exposure conditions used to be generalised, however, with a better understanding of the impact of the environment on concrete exposure, classification has also seen changes.

3.11. Concrete durability tests

Due to the wide array of factors affecting concrete durability, there currently is no single test that can be undertaken to show that a concrete is durable. Researchers have developed tests for individual concrete characteristics to make a prediction on the durability of an individual concrete. These characteristics include concrete penetrability and response to transport characteristics. Tests such as porosity measurement, permeability tests, diffusion tests and sorption tests have been developed internationally (Lamond and Pielert, 2006).

Regionally in South Africa, the South African Durability Index method has been developed. Three tests have been developed that are used to determine four concrete characteristics; Oxygen permeability index (OPI), porosity (P), water sorptivity (WS), chloride conductivity (CC). The test methods were developed from works by Ballim, Alexander and Beushausen (University of Cape Town, 2017). The basic principle of the testing regime was a realisation that improvement of concrete durability understanding rests in an ability to measure characteristics in concrete within the cover zone (Alexander, Ballim, Beushausen, 2009). These tests measure specific transport-related characteristics of concrete which can directly be linked to specific deterioration mechanism of concrete. The tests have been developed so that they can be used to assess the use of different concreting materials, mix designs and quality control on-site (University of Cape Town, 2017).

3.11.1. Oxygen Permeability Index

Oxygen Permeability test measures the pressure drop of oxygen contained in a sealed oxygen permeameter with a thin (25-30 mm) concrete sample 68mm in diameter. The laboratory test procedure for oxygen permeability index (OPI) has been specified in University of Cape Town (2017). A pressure drop indicates a loss of oxygen through the concrete slice. The test is undertaken either for a period of six hours or as long as the pressure drops from 100 KPa to 50 KPa; whichever of the two occurs first. Measurements of pressure drop are taken periodically and thus at the end of the test, the rate of pressure drop (in the Oxygen) can be calculated and also the permeability of the concrete sample. OPI is then calculated as the negative log of the permeability. In the test method, a minimum four (4) samples are used, and the OPI of the concrete is taken as the average of the 4 OPIs. OPI can be used to predict carbonation of concrete in differing environments by utilising an empirical prediction model for carbonation that utilises oxygen permeability (Ballim *et al* 2009).

3.11.2. Water sorptivity and concrete porosity

The test was developed by modifying the Kelham Sorptivity test (Ballim *et al,* 2009) and its procedure is specified in University of Cape Town (2017). In the tests a concrete specimen, 25-35 mm thick and 68mm diameter, with uniform low moisture content has one face exposed to water with the sides sealed to ensure uni-directional absorption. The specimen is weighed at regular intervals until just before it saturates, this provides the mass of water absorbed by the specimen. The specimen then is vacuum saturated in calcium hydroxide saturated water for an 18 hour period after which it is weighed to determine the vacuum saturated mass. From the test, the porosity can be determined by looking at the total water mass absorbed from start of test to the end. A graph of the mass gain against the square root of test time enables determination of water sorptivity of the specimen based on the porosity. The water sorptivity index (WSI) for the concrete then is the average of the usable water sorptivity as based on limits specified in the test.

3.11.3. Chloride conductivity

The chloride conductivity (CC) test is a type of accelerated diffusion test in which the diffusion of chloride ions through a concrete specimen occurs due to an applied potential difference on either side of the specimen. The laboratory test procedure for CC has been specified in University of Cape Town (2017). A specimen, 30 mm thin and 70 mm diameter, cut from either a cube or a core (taken from site) is preconditioned to ensure standard pore water pressure then vacuum saturated in NaCl solution. The specimen is weighed before and after vacuum saturation. The specimen is then placed in a test rig (with an anode and a cathode that's filled with 5M NaCl solution) and a DC power supply. Once the solution is filled in the cathode, DC power is adjusted with voltage and current measured until the voltage across measures 10V. The test takes a maximum of 15 minutes from the time the specimen comes out of the vacuuming machine. The chloride conductivity is found as a function of the current applied, voltage, thickness and cross-sectional area of the specimen. Chloride conductivity index of the concrete is calculated as the average of the chloride conductivity of individual specimens made from the sampled concrete (a minimum of 4 specimens are recommended). Chloride conductivity index can be correlated with chloride diffusion in concrete and can be used in prediction of chloride migration into a concrete in a specific environment.

3.11.4. Applicability of South Africa's DI method

In their research, Alexander, Ballim, Beushausen (2009) highlighted the practical applicability of the DI method. For construction works the DI method has been shown to be sensitive to material and construction effects. The tests can thus be used for design and site quality control. The practical applicability of the DI method has been tested by a number of researches, including comparing the method to other internationally developed test methods.

Beushausen and Alexander (2008) compiled comparative tests undertaken in different parts of the world to assess the effectiveness of testing quality of concrete cover using various methods, including DI method. The study showed that the chloride conductivity and oxygen permeability (from the DI method) produced results that could not only be correlated to other accepted international tests but also be used to determine cover quality by evaluating specific deterioration transport characteristics.

Alexander et al (2017) evaluated the effect of binder content, water/binder ratio and curing on concrete strength and durability affecting transport properties using the DI method. The research confirmed that increasing binder content generally reduced strength and lowered concrete durability. In the end, the research showed that to get the best out of concrete, setting concrete performance needs was far better than prescribing minimum binder content or water/binder ratio.

Alexander, Ballim, Stanish *(*2006) detailed research undertaken to assess repeatability and reproducibility of the DI method across a total of nine (9) laboratories in South Africa. This research was undertaken following previous recommendations that had highlighted the need to make test procedures user-friendly. Stanish *et al (*2006) showed that the chloride conductivity test procedures required working on as most of the laboratories produced results that showed difficulty in implementing the procedures. The other tests were found to present consistent results.

Alexander, Salvoldi and Beushausen (2015) undertook research to develop a relationship between oxygen permeability and carbon dioxide diffusion coefficient. In the research, concrete specimens (from different concrete mixes) were tested for accelerated carbonation (with varying carbonation depth) and oxygen permeability index. The results produced showed a good correlation between carbonation coefficient results and permeability of all concrete specimens. This led to the development of a regression equation correlating concrete permeability to the diffusion coefficient (generated from carbonation coefficient) independent of binder used.

Santhanam (2013), in assessing how India could implement a performance-based approach recommended using the South African DI method. The method was specifically recommended due to among others, the extensive research that has gone into developing it.

The South African National Roads Agency (SANRAL) has implemented the use of the DI method for acceptance of concrete works in bridge construction. SANRAL has provided limiting values for OPI (to inland structures) and CC (to marine structures). In Nganga *et al* 2017, the use of the DI method was assessed on various road bridge projects across South Africa for a specific period. The study concluded that the method had been successfully implemented in South African concrete practice. Table 3-1 presents the limiting values that SANRAL uses for acceptance of concrete works.

A major sign of the confidence in the test methods was when the national standard accepted the procedures as standard procedures for measurement in South Africa. In 2015, the test procedures for OPI, WSI and CC were incorporated into the national standards of testing for concrete works in South Africa as SANS 3001-CO3-1:2015, SANS 3001-CO3-2:2015, and SANS 3001-CO3-3:2015 (University of Cape Town, 2017).

3.12. Conclusion

Concrete is a heterogeneous material whose final qualities are dependent the materials used, quality of processes followed in production and early treatment it receives. Once built concrete can suffer degradation due to physical, chemical or mechanical attacks which would prevent it from full performance during its intended design life. Of all attacks corrosion of reinforcement is the single most damaging attack globally. Resistance of this degradation by concrete is dependent on the quality of the cover to reinforcement. In South Africa they have now developed tests which are able to show the quality of concrete (especially the cover concrete). The tests are also able to assist in assessing quality of construction methods and can be used as part of a site quality control management. These tests have currently been accepted as national standards for testing of concrete quality highlighting their accuracy and acceptability.

Carbonation-Induced Corrosion (from Atmospheric & Industrial)										
Designation	Description	Condition of Exposure	Description of Exposure	Typical Examples where applicable	Recommended Minimum Cover (mm)	In-situ Durability Index for various Cover Depths within Exposure Condition - 100 Year Life				
							OPI (log scale)		Sorptivity (mm/h)	
						Cover Depth (mm)	Recommended value	Minimum value	Recommended value	Maximum value
XC1a	Low hum. (<50%); exter. conc. sheltered from moisture, arid areas; interior concrete	Mild	Inland dry areas - arid to semi-arid, Karoo etc. Very low (<40%) to low humidity (40% - 50 %). Concrete surfaces not in contact with ground, protected against wetting.	Arid areas, infrequent rain: all exposed members; sides of decks & beams; deck soffits; enclosed surfaces (e.g. interior of box girders); surfaces protected by waterproof cover or permanent formwork not likely to be subjected to weathering; interior members in buildings;	40	40 mm min. cover	N/A	N/A	10.0	12.0
XC1b	Permanently wet or damp		All areas with access to external or environmental moisture Saturated conditions (RH >95%). Concrete surfaces above ground level kept permanently moist by exposure to water; concrete that never appreciably dries. Concrete surfaces below ground such as piles and buried foundations or abutments kept permanently damp.	Partially submerged and hydraulic structures kept permanently damp; drainage & other elements kept moist; surfaces in contact with permanently damp soil; surfaces kept damp by condensation or moisture; piles (both dry cast and against casings)	40	40	9.20	9.00	10.0	12.0
						50	9.00	9.00	10.0	12.0
						60	n/a	n/a	n/a	n/a
XC2	Wet, rarely dry	Moderate	All areas with access to external or environmental moisture Concrete surfaces above ground level kept mostly in moist condition by exposure to water; concrete may occasionally dry for appreciable periods such as when tanks are emptied	Partially submerged and hydraulic or drainage structures kept mostly damp; surfaces in contact with mostly damp soil; surfaces kept mostly damp by condensation or moisture; all wet or mostly damp surfaces which may occasionally dry for limited periods	40	40	9.40	9.00	10.0	12.0
						50	9.10	9.00	10.0	12.0
						60*	9.00	9.00	10.0	12.0
						70*	n/a	n/a	n/a	n/a
XC3	Moderate Hum. (50-80%). Ext. conc. sheltered from rain in non-arid areas		Near-coastal areas with no chlorides; moist inland areas; adjacent to dams, lakes, major rivers Moderate humidity (50% to 80%), moist climate. Exterior concrete surfaces in moist areas or adjacent to major water bodies, permanently sheltered from rain or direct surface moisture	Moist areas: sides of beams protected from direct rain; deck soffits; enclosed surfaces (e.g. interior of box girders); surfaces protected by waterproof cover or permanent formwork not likely to be subjected to weathering. Consider additional cover at edges of deck at expansion joints, soffits of cantilevers and	40	40	9.40	9.00	10.0	11.0
						50	9.10	9.00	10.0	11.0
						60*	9.00	9.00	10.0	11.0
						70*	n/a	n/a	n/a	n/a
XC4	Cyclic wet and dry	Severe	All areas with access to external or environmental moisture; arid areas excluded Moderate humidity (50% to 80%), moist climate. Concrete surfaces exposed to rain or alternately wet and dry conditions	All exterior surfaces exposed to rain; surfaces where heavy condensation takes place; surfaces alternately wetted and dried by drainage or environmental moisture, such that moisture may penetrate concrete member.	45	40	9.60	9.20	10.0	10.0
						50	9.30	9.00	10.0	10.0
						60*	9.10	9.00	10.0	10.0
						70*	9.00	9.00	10.0	10.0

WARNING: Covers shown with a asterisk (*) should be avoided so as to (i) limit crack widths, and (ii) ensure durability concrete is being specified and must be discussed with the client before being specified. NOTE: Heavily Polluted Industrial Areas : Increase Cover for any exposure Condition above by 10mm

Chloride-Induced Corrosion (from Groundwater, Seawater & Sea spray)												
Designation	Description	Condition of Exposure	Description of Exposure	Typical Examples where applicable	Recommended Minimum Cover (mm)	In-situ Durability Index for various Cover Depths within Exposure Condition - 100 Year Life						
						Cover Depth (mm)	Chloride Conductivity (mS/cm)				Sorptivity (mm/h)	
							Typical Binder Blends					
							70:30 CEMI:FA	50:50 CEMI:GGBS	50:50 CEMI:GGCS	90:10 CEMI : CSF	Recommended value	Maximum value
XS1	Exposed to airborne salt but not in direct contact with seawater or inland saline waters	Very Severe	Proven presence of chlorides; generally < 1km from sea, and coastal river valleys (where chlorides are present) and estuaries, or the presence of chlorides proven by experience or testing. This will include inland salt pans or groundwater carrying slats, etc	All exposed and external surfaces subject to significant airborne salt, any surface on which salt can deposit from the air	50	40	1.50	1.60	2.10	0.40	10.0	12.0
						50	2.10	2.20	2.80	0.50	10.0	12.0
						60	2.60	2.70	3.40	0.65	10.0	12.0
XS2a	Permanently submerged in sea (or saline waters)	Severe	Permanently (or substantially) submerged in the sea (without heavy wave action), in coastal saline estuaries & rivers, in any aggressive saline waters Concrete surfaces exposed to heavily polluted industrial waters, permanently or substantially submerged or permanently wet saline conditions (Generally oxygen starved area approximately 1-1,5m below spring type level)	Coastal or other structures permanently submerged in seawater or other aggressive saline waters, including industrially polluted water; surfaces of structures in contact with marshy conditions	50	40	1.00	1.10	1.40	0.30	10.0	11.0
						60	1.40	1.60	2.00	0.40	10.0	11.0
						60	1.80	2.10	2.50	0.50	10.0	11.0
XS2b	XS2a + exposed to abrasion	Extreme	As above, but with heavy wave action, in any aggressive saline waters where abrasion occurs	As above + exposed to abrasion	60 (Mandatory)	60	1.45	1.70	2.00	0.40	10.0	11.0
XS3a	Tidal, splash & spray zones		Sea or saline estuaries and rivers, but not permanently submergec; tidal zone; and in a spray or splash zone. surfaces exposed to aggressive saline waters, including heavily polluted industrial waters, without being permanently wet.	Coastal or other structures exposed to intertidal, splash, or spray zones, or exposed to other aggressive saline waters, including industrially polluted waters, without being permanently wet; members subject to burying by aeolian sands near coast	50	40	0.65	0.85	1.00	0.25	10.0	10.0
						50	1.10	1.35	1.45	0.35	10.0	10.0
						60	1.45	1.70	2.00	0.40	10.0	10.0
XS3b	XS3a + exposed to abrasion	Extreme	As above, but with heavy wave action or where abrasion or erosion can occur	As above + exposed to abrasion	60 (Mandatory)	60	1.10	1.30	1.55	0.30	10.0	10.0

Notes:

1. Exposure Classes

i). Exposure classes are only best estimates at this stage and considerably more work is needed on this.
ii) The key to interpreting the exposure classes is that the steel should 'feel' the impact of the exposure. E.g. wetting and drying should really influence the concrete at the level of the steel, rather than being a fleeting surface wetting.
iii) Various bridge elements will experience the same exposure class in different ways. E.g. interior columns and deck undersides will generally remain dry, while deck edges, exposed abutments, and balustrades will experience the full climatic effects.

2. Cover:

i) Minimum cover for bridge structures is taken as 40 mm, i.e. civil engineering structures are contemplated.
ii) In-situ piles shall in general have cover not less than 75mm due to tolerance variation
iv) Variable cover should be considered for bridge design:
- Cantilevers and balustrades
- Soffits and interior columns

iii) Pre-cast piles shall not be lesser than than 55mm

3. OPI

i) Values are based on UCT spreadsheets.
ii) Most values are based on a blended binder, not a pure OPC binder.
iii) UCT's spreadsheet tends not to differentiate between OPC and Slag mixes, but does show more conservative values for FA mixes. The values in the spreadsheet tend towards the FA mix values, since a great deal of concrete in South Africa, particularly the interior regions, contains FA.
iv) The justification for the above is that it is not possible to always know what binders will be used in construction concretes, and therefore a conservative approach is justified.

4. Chloride Conductivity

i) Values are based on UCT spreadsheets.

ii) In this case, allowance is made for the different binder types.
iii) Interpolation or extrapolation of the CC values taken from UCT spreadsheets for the different exposure classes

4. Prescriptive concrete durability design and specification, Namibian perspective

4.1. Introduction

Traditionally, the design and construction of concrete has followed strict limits which are set in standards and specifications. These limits have come from years of experience and in a number of instances with little scientific backing on the relationship of the limits and the performance of concrete (Lemey, Lobo, Obla 2006). The limits have covered all aspects of works from design limits, material limits, and construction method limits. In Namibia, concrete designs have generally followed South African design codes, mostly the now replaced SABS 0100-1. This code is based on BS 8110, as highlighted in its introduction. SANS 10100 which has replaced SABS 0100-1 is slowly being adopted in design offices. For construction specification, a number of different specifications are used. The two commonly used are SANS 1200 (standardized specification for civil engineering construction) and Standard Specification for Road and Bridge for State Road Authorities (produced by South Africa's Committee of Land Transport Officials 1998 edition). Along the coast, NAMPORT Technical specifications are used by the port authorities. This chapter highlights some prescriptions that the codes and specifications impose on designers, producers and builders of concrete works which are implemented in Namibia's concrete construction industry.

4.2. Design

4.2.1. Basis of design

In BS 8110-2:1 the code emphasizes the fact that design according to the code looks at safety, serviceability and durability during the design life of a structure. The basis of design introduces the limit states to which concrete will be designed to. Two limit states are introduced: ultimate limit state and serviceability limit state.

Ultimate limit state governs the safety of a design, ensuring that there is a high probability of the structure (or structural element) not failing during its design life. Thus ultimate limit state consideration is made for stability, robustness and special hazards (for specific structures). In BS 8110-2.4, guidance is provided for materials use to comply with the design code specification. Serviceability limit state governs the usability of a designed structure (or structural element), aiming to ensure that a structure will behave satisfactorily under working conditions. The limit state ensures that the structure will not have excessive deflection or vibration or cracking.

Beeby et al (1987) highlight that the code was prepared with the belief that "knowledge was not adequate to design concrete structures to specific durability and life". This implies the durability service life of concrete structures, according to the code was not analytically established.

4.2.2. Durability

BS 8110-2.2.4 highlights the need for durability considerations and raises a number of points worthy of being considered in the design. These include exposure conditions, cover to reinforcement, concrete/steel protection, material selection and mixing, and workmanship.

4.2.3. Exposure

The environment in which concrete is cast is defined according to BS 8110-3.3.4, and the exposure conditions are provided in Table 4-1

Table 4-1: Exposure condition for reinforced concrete design (BS 8110)

Environment	**Exposure Condition**
Mild	Concrete surface protected against weather or aggressive conditions
Moderate	Exposed concrete surfaces but sheltered from severe rain or freezing whilst wet Concrete surface continuously under non-aggressive water Concrete in contact with non-aggressive soil-water (see sulfate class 1 of Table 7a in BS 5328-1:1997) Concrete subject to condensation
Severe	Concrete surface exposed to severe rain, alternate wetting and drying or occasional freezing or severe condensation.
Very severe	Concrete surface occasionally exposed to seawater spray or de-icing salts (directly or indirectly) Concrete surface exposed to corrosive fumes or severe freezing conditions whilst wet
Most severe	Concrete surface frequently exposed to seawater spray or de-icing salts (directly or indirectly) Concrete in seawater tidal zone down to 1 m below water
Abrasive	Concrete surface exposed to abrasive action e.g. machinery, metal tyred vehicles or water carrying solids.
NOTE 1 For aggressive soil and water conditions see 5.3.4 of BS 5328-1:1997 NOTE 2 For marine conditions see also BS 6349	
*For flooring see BS 8204	

The exposure classifications highlight the different conditions that concrete can be cast in. It will be appreciated that the exposure classification as presented is not directly linked to concrete deterioration but environmental aggressiveness. In this case, the responsibility is on the designer based on his/her experience and knowledge to assign the exposure class.

4.2.3.1. Concreting materials

Beeby et al (1987) highlight that in concrete design (based on BS 8110) characteristic concrete strength is used as limit state is based on structural deformation. This would, in part, explain why most construction specifications have adopted concrete strength as an acceptance criterion. BS 8110 has provided minimum characteristic strength based on the concrete cover provided in specific environmental condition. In this, the code highlights the deterioration problem of concrete in different environments. Clause 2.4.7 specifically links durability to design, specification and construction. However, clause 2.4.6 refers to clause 3.3 and Table 3.4 of the code (presented as Table 4-2) which prescribes material use for defined exposure condition. In this case, the minimum strength grade, minimum water/cement ratio and minimum cement content are set by the code.

Table 4-2: Nominal concrete cover for concrete design (BS 8110)

Nominal cover to all reinforcement (including links) to meet durability requirements					
Condition of Exposure	**Nominal cover (mm)**				
Mild	25	20	20*	20*	20*
Moderate	-	35	30	25	20
Severe	-	-	40	30	25
Very severe	-	-	50+	40+	30
Extreme	-	-	-	60+	50
Maximum free water/cement ratio	0.65	0.60	0.55	0.50	0.45
Minimum cement content (Kg/m^3)	275	300	325	350	400
Lowest grade of concrete	C30	C35	C40	C45	C50
*These covers may be reduced to 15 mm provided that the nominal maximum size of aggregate does not exceed 15 mm					
+Where concrete is subject to freezing whilst wet, air entrainment should be used (see 3.3.4.2)					
NOTE 1. This table relates to nominal aggregate of 20mm nominal maximum size					
NOTE 2. For concrete used in foundation to low rise buildings, see 6.2.4.1					

Acknowledgement to the possibility of other mix designs using cementitious materials other than Portland cement is made in Clause 6.1 of the code. Clause 6.2 has further placed limitations on the mix proportions, minimum cement/water ratio, and minimum strength for durability consideration. Compliance to the required strength, in this case, is deemed to imply concrete will be durable and thus perform in its environment. For concrete with ground granulated blast furnace slag and fly ash, the code places stricter limitation due to lack of knowledge on the performance of these concretes (Beeby et al, 1987).

With regards to aggregate, the code requires the performance of the aggregate in concrete to be the dominant factor for the choice of aggregate to be used. In clause 6.1.3.2, the code emphasizes that performance, in this case, refers to the behaviour of the aggregate with respect to fresh concrete properties (workability and placement). There is no mention of the performance effect on hardened concrete.

The benefits of admixtures in concrete are highlighted by the code though their performance is tied into the strength attainment of the concrete. With durability consideration, there is a limitation placed on the total chloride in concrete (clause 6.1.5.4).

4.2.3.2. Concrete cover

In the BS 8110, the minimum cover is provided mindful of workmanship requirements (bar size placement, and material consistency), fire and exposure consideration. Table 4-2 has provided for minimum cover based on exposure condition and in Fig. 4-3 minimum cover with respect to fire resistance is provided.

Table. 4-3: Minimum cover based on fire resistance (Source: BS 8110)

Nominal cover to all reinforcement (including links) to meet specified period of fire resistance (see note 1 and 2)							
Fire resistance	Nominal cover						
	Beams		Floors		Ribs		Columns
	Simply supported	Continuous	Simply supported	Continuous	Simply supported	Continuous	
h	mm	mm	mm	mm	mm	mm	mm
0.5	20+	20+	20+	20+	20+	20+	20+
1	20+	20+	20	20	20	20+	20+
1.5	20	20+	25	20	35	20	20
2	40	30	35	25	45	35	25
3	60	40	45	35	55	45	25
4	70	50	55	45	65	55	25
*For the purpose of assessing a nominal cover for beams and columns, the cover to main bars which have been obtained from Table 4.2 and 4.3 of BS 8110: Part 2: 1985 have been reduced by a notional allowance for stirrups of 10mm to cover the range 8 mm to 12 mm (see also 3.3.6							
+These covers may be reduced to 15mm provided that the nominal maximum size of aggregate does not exceed 15mm (see 3.3.1.3)							
Note 1. The nominal covers given relate specifically to the minimum member dimension given in Figure 3.2 Guidance on increased cover's necessary if smaller members are used in section four BS 8110: Part 2: 1985							
Note 2. Cases that lie below the bold lines require attention to the additional measures necessary to reduce the risk of spalling (see section four of BS 8110: Part 2: 1985)							

4.2.4. Concrete handling, placement and curing

The code does not place any limitation with regards to handling and placement of concrete. It rather, in Clause 6.5, highlights the need to deliver concrete quickly to site (after casting) and close to the intended placing spot. The code recognizes the different roles of concrete producers and those placing on-site. For cast concrete, the code in Clause 6.6 requires a minimum period of curing that is linked to the cement used in the concrete. Further on, clause 6.9, provides minimum striking time for shutters, and it acknowledges possibilities of using accelerated curing methods.

4.3. Construction works specification

4.3.1. Standardized specification for civil engineering construction (SANS 1200)

SANS 1200 is a South African Bureau of Standards specification for undertaking civil engineering construction work. Within the suite, the main document for concrete works is SANS 1200 G. Specialized works such as precast, prestressed and piling are not covered under this part of the specification (Clause 1.2).

SANS 1200 G Clause 2.4 provides for exposure classification for concrete works which is also applicable to all the other specialised works. In this classification, it is noted that the exposure condition is more on the general action of the environment on any concrete placed within it. The specification in Clause 2.4 provides minimum cover specification, which is based on these exposure classifications and no deviation from these minimum covers is allowed in Clause 6.2.3(e). It is interesting though that in spite of not allowing for deviation, there is no requirement for testing of compliance for this cover provision.

SANS 1200 G Clause 5.5.1.5 highlights durability requirements by prescribing maximum limits of water/cement ratio. The limits are based on cross-sectional size and exposure classifications in Clause 2.4. For fresh concrete, the specification highlights the need for consistency and workability with slump being the only test demanded in Clause 5.5.1.2. Clause 5.5.1.4 sets a limit on the amount of chloride in cement just as a limitation is placed on the quality of water (in Clause 3.3) which indicates an awareness of the concrete durability needs. In SANS1200GF, for prestressed concrete, more consideration for durability is highlighted in Clauses 5.5.1 and 5.6. In both of these provisions are made for ensuring corrosion protection to the tendons.

The main criteria for acceptance of concrete (in the specification) is strength as highlighted in Clause 5.5.1 and Clause 7.3. For the specialised works, load carrying capacity has been added as an acceptance criterion. There is a deviation where prescribed mixes, as provided in Clause 5.5.1.6, are used. Following the prescribed mixes is deemed to be a compliance. The importance of concrete curing is emphasized by the provisions of minimum curing periods in Clause 5.5.8.

4.3.2. **Standard specifications for road and bridge works for state road authorities (COLTO)**

The standard specifications for road and bridge works for state road authorities (1998 edition) produced by the Department of Transport in South Africa are used in Namibian road works. The specifications cover all aspects of civil engineering works involved in highway construction including concrete. A number of series within the specification relate to concrete work with 6400 being the main one.

Series 6400 covers all compliance requirements for concrete including materials, batching, transportation, placement and testing. 6404 confirms that the main criteria for acceptance of concrete works are attainment of prescribed strength or use of prescribed mixes in line with Series 6414. Series 6404(b) dictates that where for durability reasons there is need of limiting cementitious materials and water/cement ratio, the concrete class should specifically indicate this with a prefix to the strength classification. Table 6404/2 provides prescribed nominal concrete mixes that a contractor can be asked to use. The specification does not state under what conditions these nominal mixes are supposed to be used, and compliance is purely by following the mix.

Series 6400 also provides limitations for selecting constituent materials for concrete. Consideration for durability is seen in the limitation for total alkaline content in concrete, sulphates and chloride. The specification does not set a limit for water/cement ratio except in prescribed mix case, however, it provides limitation on slump as a measure of concrete workability based on the type of concrete being produced (Table 6404/2). To ensure protection to reinforcement, Clause 6307 provides minimum cover to reinforcement in concrete. The provision of the cover here is based on environmental action on the concrete surface. The provision has further given direction on which structural elements specific environmental actions are applicable to including consideration for surface coating. The specification does not provide for checking of cover on concrete after concrete has been cast. Concrete curing is specified under clause 6409, and clause 6410 provides caution for extreme weather concreting conditions.

4.3.3. NAMPORT marine concrete specification

NAMPORT is an entity owned by the government of the Republic of Namibia with a mandate to manage all ports and lighthouses for the country. In undertaking this responsibility, NAMPORT has developed several management documents, one of which is a specification for construction works, including concrete construction. The NAMPORT technical specifications have been split up for the various works with specific provision for marine concrete, foundations, reinforcement and formwork. These specifications are used for all construction works under NAMPORT management.

The marine concrete specifications have been developed with design reference to BS 8110 and material reference to a number of British and South African standards (Clause 2.2.2). The specification of constituent materials limitation has been placed on the total content of sulphate, reactive alkaline and chloride. Clause 2.2.1 limits choice of cement for concrete works by specifically listing acceptable cement types.

Clause 4.2 indicates the liberty contractors have to undertake mix design. However, the liberty to undertake the mix design is limited by the provision of Table 1 and 2, which specify concrete strength class and concrete mixes. In the prescribed mixes, maximum water/cement ratio is provided with limitation on using CEM I cement within set quantity proportions. Acceptable mix design is required to be tested for not only strength but also chloride conductivity and oxygen permeability in line with Durability Index Tests and should satisfy the limits for the two.

For bored piles, the specifications further provide a minimum concrete strength grading, and cementitious material quantity (minimum and maximum). Performance requirement of concrete used for piling has been provided in Clause 3.4 of the bored pile specification. In Appendix C (of the specification) presentation of tests to be undertaken has been provided; however, there is no provision on when compliance with the performance requirements will be shown or presented. This becomes an issue when it is reconciled with the marine concrete specification, which only requests compliance during mix design. During construction, the current specification does not require verification of the concrete durability parameters, and thus NAMPORT has no way of confirming what has been built unless a deviation is made from the specification.

Slump testing of concrete during casting is a requirement under the specifications. Whereas no limits are given for marine concrete, under the bored piles, a limit is provided for in-tremie placed concrete. Strength testing of sampled concrete cubes has also been provided for in the specification. The specification has specified a minimum cover of 75mm for concrete works. The only provision for checking the cover is before concrete casting; there is no requirement for any cover check after casting. Concrete curing importance is highlighted in Clause 4.10 (Marine concrete specification). The specification does not allow for use of any curing compound.

4.4. Challenges of prescriptive durability design and specification

Over time, the use of prescriptive design and specifications (as seen in this chapter) has been found to have some inherent problems. Some of these problems have included:

1) Correlation between service life and concrete quality is not established – in the prescriptive approach there is no test put in place to assess whether what has been built will last the intended service life (Edvardsen 2010). In the case of Namibia, during design, there is no testing of materials or construction method. Just before commencement of concreting a regime of testing individual materials for suitability in concrete is undertaken, including concrete mix design. On casting of concrete, a sample is taken from which concrete cubes are made and cured in a controlled environment to be tested for strength. Once the cubes pass a minimum strength, the concrete is deemed to be fit. In this scenario, thus, there is

no real test that looks at the cast concrete to give quantitative proof of the quality of the concrete in relation to the design life of the structure.

2) Stifling of innovation – as construction works these days are mostly undertaken by companies that are out to make profit companies compete on cutting unnecessary costs, and concrete research is seen as an unnecessary cost. As the specification dictates what it deems a contractor needs to satisfy, there is thus no incentive to innovate on those designing, producing and erecting concrete structures (Bickley, Hooton, Hover 2006).
3) High possibility of increased cost for infrastructure development for owners – due to the lack of innovation (in producing and construction of concrete works) it implies owners of infrastructure works may have to pay more for the cost of concrete during construction as cost-effective options were not explored by the designing and construction team. Also, during operation of the infrastructure, the cost would be high if maintenance needs come in much sooner due to poor concrete durability (Lemey, Lobo, Obla, 2006).
4) The British standard was developed for the design and construction of concrete in Britain though it has been adapted for use in other countries like Namibia. There are very contrasting environmental conditions between temperate Britain and mostly semi-arid Namibia. Thus the prescription that has been placed in the code might not work perfectly for the Namibian condition. The differing geographic locations imply the definition of exposure condition of concrete might not be appropriate (Alexander, 2016).

4.5. Conclusion

In Namibia, concrete design and construction specification generally follow international prescriptive approach as the country has no local standards in this regard. NAMPORT is one institution that has slightly different specification for concrete works, however, within them there still is prescriptive requirements to concrete production. Under this approach durability is deemed to be achieved if concrete achieves a specific strength with limitation of water/binder ratio. The problem of this approach is that concrete produced is at no time checked against the environmental aggressiveness. International research has shown shortcomings of this approach of building and managing concrete structures. In the long term structures are found not to perform fully for the intended design life and there is also the limitation on innovation which has direct cost implication for both developers and contractors. Concrete structures that deteriorate quicker end up not offering the function they were built for and indirectly affect users of the structure.

5. Performance-based durability design and specification

5.1. Introduction

There has been a general concern regarding the early development of defects in concrete, even where there has been a strict adherence to prescribed design standards and construction specification (Siemens and Edvardsen, 1999). Through various research, it has come to be appreciated how the various ingredients of concrete, the environment concrete is built-in, and the construction methods affect concrete's ability to deliver acceptable performance. The traditional approach of concrete design and construction specification attempted to achieve durability of concrete by prescribing specific concrete mixing and construction parameters. However, besides attainment of the set limits, the codes have not required verification of specific performance criteria during the service life of the structure or structural element. Codes were drawn up like this because there was minimal knowledge on concrete durability and, more importantly, there was a lack of means to predict a required performance. With advances in research, there is now a better understanding of material properties and interaction between concrete and the environment. In the process, several models have been developed (and test methods) that assist in predicting the service life of structures.

5.2. Service Life

ISO 16204 previously defined service life as the assumed period for which a structure or part of it is used for its intended purpose with anticipated maintenance, but without major repair being necessary. In this definition, it is acknowledged that a structure will not exist forever but will suffer deterioration over time once built. The definition also highlights that the deterioration does not lead to the collapse of a structure, but the structure could be restored with repairs performance. This definition of service life was found to be qualitative in describing performance of structures; however, to enable it be used for quantitative design *fib* and ISO revised the definition (Helland 2017). In *fib* 2010 it is emphasised that defined service life should include: use of materials with satisfactory properties, defined appropriate limit states of durability, verification with regards to the probability of exceeding the limit states, good conceptual design, appropriate construction methods and a plan for life cycle management.

There have been a lot of structures that have deteriorated much faster than anticipated and to ensure the structures continue to perform up to their intended service life, maintenance has been undertaken. This was graphically presented by Alexander, Ballim, Beushausen *(*2009) and is shown in Fig. 5-1. To minimise early maintenance, performance-based design is now being recommended.

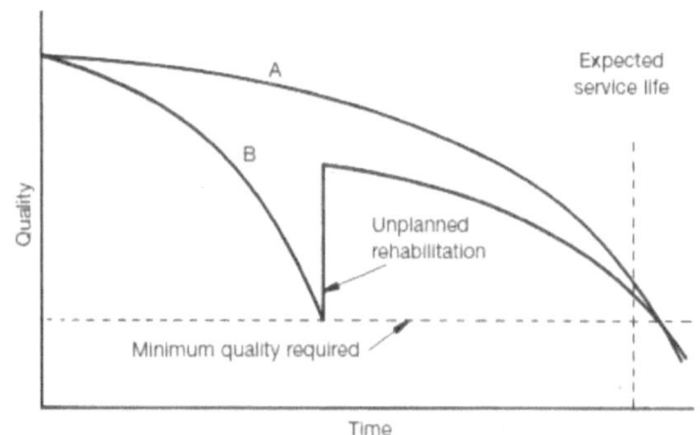

Fig. 5-1 : Structural rehabilitation and deterioration scenarios (Source: Alexander, Ballim, Beushausen, 2009)

5.3. Performance-based durability design and specification

In performance-based design, the required performance of a structure (or structural element) is predetermined with a maximum probability at the end of the design life (Breugel, Polder, Wegen, 2012). Thus the design is a probabilistic design concept that uses the influence of the environment on a structure over time. Mathematical models of the deterioration mechanisms are then used to determine the service life of the structure (or structural element).

The principles of performance-based design for concrete structures are summarised in Fig. 5-2.

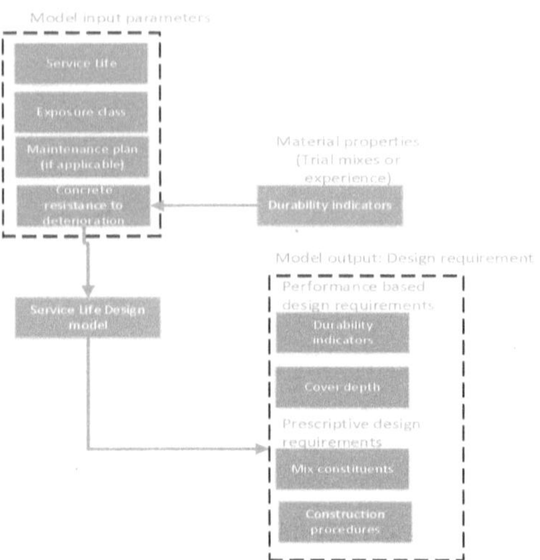

Fig. 5-2: Principles of performance-based design of concrete structures (Source: Alexander *et al,* 2015)

5.3.1. Exposure classification

In performance design and construction specification, concrete is designed and built to perform in a specific environment. It is thus important to properly define the environment that concrete will be exposed to. With a better understanding of deterioration mechanisms of concrete, the definition of concrete exposure has been changed.

The Eurocode 2 (EN206-1:2000) has provided a classification of exposure based on anticipated deterioration mechanisms and severity. In general corrosion, freeze/thaw and chemical attack have been identified as the deterioration mechanisms against which concrete should withstand. Corrosion has further been split into two distinct classifications: one for carbon-induced corrosion and another for chloride-induced corrosion. This classification ensures that concrete designed is specific to the 'demands' of the environment it is exposed to. The American code for concrete design, ACI 318, has a somewhat similar classification with a major difference being the classification of corrosion. It looks at protection rather than the cause of the corrosion, and thus the classification is not as extensive as the one provided by EN206 1:2000. In general, these (EN206-1 and ACI 318) design guidelines have shown that in defining exposure of concrete, more interest should be on the actual deterioration and the environment than the environment only (independent of the concrete).

EN 206-1 has been framed such that it allows different countries to extend the provision in code to suit specific national conditions. BS 8500-1:2006 has extended the environmental exposure classification beyond provisions of EN 206 1:2000 (a summary of concrete exposure is presented as Table 6-2). Alexander (2016) explained how South Africa is in the process of adopting EN206-1 for concrete design through the inclusion of a national guidance document.

5.3.2. Service life design models

There are a number of service life design models currently being used. Current models have used corrosion (of reinforcement) as a measure of performance due to the dominant deterioration effect of reinforcement corrosion on concrete structures. DuraCrete, developed in Europe, models chloride ingress into concrete and carbonation. There is also the Scandinavian model "ClinConc" and the South African "chloride- and carbonation-induced corrosion initiation models".

Performance-design ensures optimization of concrete mix (and mix constituents selection) as the testing of materials is undertaken during the design phase when the 'true' capability of the materials regarding deterioration is assessed and documented. During construction, quality control becomes a process of controlling the variation of material parameters and geometric variables (Siemes, 1999). Thus tests, during construction become more of a compliance checks.

Edvardsen (2010) presented a project that led to the development of the European service life model, DuraCrete, which was the forerunner in the development of performance-based design models. The initial model was a fully probabilistic approach that modelled durability based on limit state design principles. In this case, the environmental exposure on concrete over time becomes the loading, and the resistance is offered by the concrete quality. The full probabilistic approach requires development of validated probabilistic models, quantifiable parameters and reproducible test methods for a design (ISO 16204). This becomes unrealistically uneconomical and is thus limited to use in exceptional structures. To this end, *fib* bulletin 34 proposed three other approaches that can be used with the full probabilistic approach restricted to special structures needing long service life. A deterministic approach was recommended, which uses factors of safety to consider the environmental effects and material resistance. A third approach is deemed-to-satisfy option. In this case, the rules will be based on calibration of the fully probabilistic models. Finally, an approach that completely avoids the deterioration is acknowledged. This is done by the use of non-reactive materials. The fib bulletin 34 design approach is illustrated in Fig. 5-3

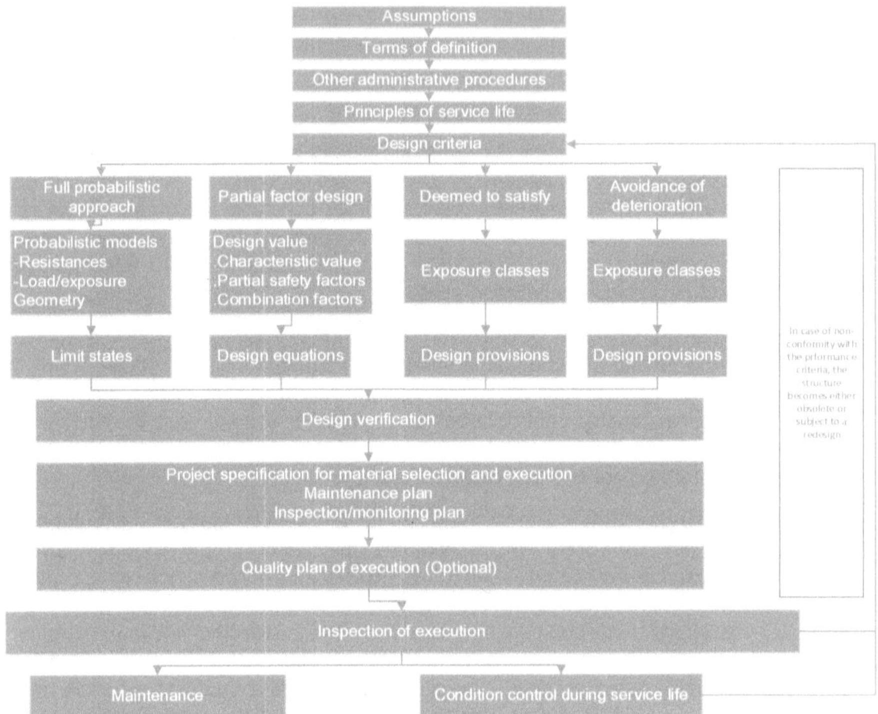

Fig. 5.3: Flow chart "Service life design" (Source – Fib bulletin 34)

5.4. Developing a performance-based design and construction specification approach

When developing a new design and construction specification approach, it is important to understand critical basic requirements to ensure effectiveness. Siemes and Edvardsen (1999) highlighted some of the key issues for a performance-based system

a) Performance definition – this is where the owners define the performance of the structure and related service life. In this particular case, the service life definition with regards to reinforcement corrosion in the structure determines performance. The owners then should determine when a service failure occurs and when ultimate failure will be deemed to have occurred. This will be in line with the 3-phases corrosion damage model (Fig. 4-4).

There is a varied definition with regards to what is meant by the end of service life in concrete structures. Helland (2017) highlighted this in a presentation where he proposed the use of a unified definition of the end of service life. In Europe most countries use depassivation as the end of service life (with varying degrees of depassivation) whilst others use cracking and spalling as the end of service life. In South Africa, Alexander, Ballim, Beushausen (2009) indicated that the time taken to depassivation is the service life.

b) Identification of relevant degradation model – as mentioned, various degradation models are currently available. The models enable the evaluation of changes over time in material and structure based on the environmental conditions and concrete mixes. An appropriate model to be followed should be chosen that is supported by technical capacity and research.

c) Testing and quality control at various stages of a structure's life – the performance framework calls for testing of concrete at the design stage, production, placement, and when it is in use. The purposes of testing will be different at different project stages. At design, one is looking at providing a basis for controlling material and geometric variation. Compliance tests at production stage are to show adherence to the design requirements. In use, the concrete is tested for verification purposes realising the influence placement has on quality of concrete to resist degradation. Alexander, Beushausen, Torrent (2008) emphasised the issue of quality control and testing by stating that most of the premature concrete deterioration (related to corrosion) is a product of minimal understanding of factors affecting durability as well as inadequate means of enforcing or testing compliance.

5.5. Performance-based design and specification implementation challenges

Performance-based durability design and specifications have been implemented in a number of countries and challenges faced by them have been documented. These challenges provide guidance for any other country looking into implementing performance-based design and specification. Challenges faced have included the following:

a) Legislative conflict – in New Zealand, when performance-based practices were introduced, they clashed with deregulation and introduction of new codes (Buchanan, Gibson, Morris, 2006). It is thus important that having understood the industry needs from a technical point, the various legislative frameworks that impact on concreting should be understood. This should include any new coming changes which would impact on the implementation of a new concreting approach.

b) Practitioner resistance – due to the entrenched use of prescriptive methods, a number of projects have encountered challenges to convince practitioners to embrace performance-based methods (Camarini *et al,* 2010). This has mostly been due to the fact that the new regime demands more (or better) quality verification than the previous ones. Also, due to the nature and size of project, practitioners are not very willing to implement performance-based approaches on projects which are routinely undertaken.

c) Lack of local service life models – this goes to the issue of research at a local level to ensure that any adopted models are developed with local environmental conditions. This was a specific problem that was faced in Brazil (Camarini *et al*, 2010). This challenge becomes the same as the accusation that is made against codes that are developed for other countries and used in another.
d) Lack of knowledge – a knowledge of the deterioration mechanism is vital to ensuring appreciation of the value that performance design and specification bring to the whole building life cycle cost.

5.6. Conclusion

International research has shown that concrete is directly affected by the environment it is built in. This has led to development of durability design, construction and management of concrete that is based on how concrete performs in its environment. To be able to undertake this approach nations have developed national models based on the most influential deterioration mechanism, corrosion. There is no single uniform model rather based on capabilities countries have developed their own models. Using these models engineers are able to determine the appropriate material mix for a specific environmental aggressiveness and once works commence the models are used to ensure compliance with quality requirements. Once the structures are built the models can be used to verify the concrete built. As good as this approach sound implementing it in a ny place has its own challenges. The approach requires a buy-back from industry to be able to be implemented. Not only should industry accept it but they need specific training in concrete deterioration and performance based approach.

6. Project area concrete condition assessment overview

6.1. Introduction

Chapter 2 highlighted the varying environmental conditions that are prevalent in Namibia, which present different challenges to concrete structures built across the country. A visual assessment was undertaken, to assess the concrete conditions, in four towns in Namibia within a 250 Km stretch: viz Walvis Bay, Swakopmund, Arandis and Usakos. Walvis Bay and Swakopmund are 30 km apart, located along the Atlantic coast, whereas Arandis and Usakos are 50 km and 140 km from the Atlantic coast (respectively). The four towns are shown in Fig. 6-1.

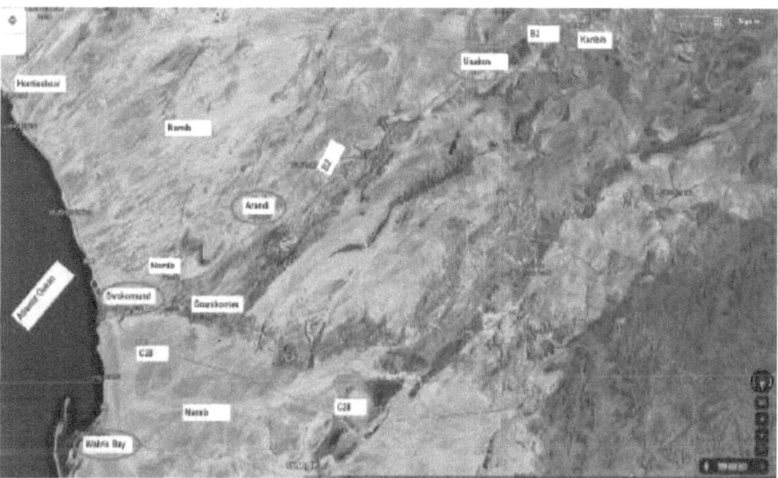

Fig. 6-1 – Sketch map showing location of the research towns are circled in red (Source: Google, 2017)

The towns chosen, for visual assessment of concrete structures, were grouped into two; arid and coastal towns. For each climatic condition two towns were selected so that a comparison could also be undertaken within similar environmental condition. The choice of which infrastructure to undertake visual assessment on was based on the following

a) the ability to engage infrastructure owners
b) traceability of design standards and construction specification for similar works currently being implemented
c) accessibility of infrastructure

Based on the 3 criteria set out 4 different infrastructure groups were chosen for assessment. The total selections from the groups are presented in Table 6-1.

Table 6-1: Visual inspection sample

	Arid zone		Coastal zone		Management institution
	Usakos	Arandis	Swakopmund	Arandis	
Bridges	3	3	2	2	Roads Authority
Marine	*	*	1	2	NAMPORT
Sewer Treatment	**	**	1	**	Local municipality
Water tanks	2	**	2	2	NAMWATER

* Not available

**Inaccessible

6.2. Walvis Bay

6.2.1. Marine structures

The Walvis Bay port was built around 1960. Within the port, there are many reinforced concrete structures built to enable handling of ships for different purposes. The port has two main structures for marine vessel use: the small craft jetty and the main port quay. Due to accessibility issues, only the small craft jetty was inspected.

6.2.1.1. Small craft jetty

The small craft jetty was built for handling small vessels including tug boats and rescue vessels. The jetty structure has concrete piles carrying concrete beams and slabs. The superstructure of the jetty gets variedly submerged in water based on prevailing tide conditions. Vessels dock against improvised rubber fenders made of used rubber tyres which are hung from the concrete superstructure (Fig. 6-2).

Fig. 6-2: Part of the small craft jetty structure

Port management highlighted incidents of vessels directly striking the concrete (due to loss of the improvised fenders) that led to chipped edges of the superstructure. During the inspection, several concrete defects were identified (Fig. 6-3). The worst of these cases had concrete spalling due to corrosion of the embedded reinforcement. Surface material loss was also observed on the top surface of the jetty structure (which is a trafficked area). This material loss would reduce the cover provided to reinforcement and, in the process, increase the risk to reinforcement corrosion.

(a) Concrete spalling

(b) concrete scabbling

(c) Expansion joint concrete disintegration

(d) concrete spalling

(e) Surface material loss

(f) reinforced concrete repair works

Fig. 6-3: Small craft jetty and concrete deterioration

To fix the damages, part of the jetty was cordoned off to allow for repairs of the concrete structure. With the closure, the port suffers financial loss from inability to handle ships in addition to the repair costs.

6.2.1.2. *Langstrand jetty*

An amusement area has been developed between Walvis Bay and Swakopmund at a beach called Langstrand. Amongst the many facilities provided is a jetty structure with an entertainment building at the end and man-made pools that fill with overflow from the ocean (Fig. 6-4). The jetty structure is a timber structure carried on stub concrete columns of varying heights. The pools have a concrete wall on the western edge bordering the Atlantic Ocean, and they fill when a wave strikes against the wall and spills into the pool. A concrete retaining wall has been built on the southern edge of the jetty structure to prevent sand build up into the pools.

Fig. 6-4: Langstrand Jetty and municipal pools

The retaining wall was found to have developed extensive cracking. No sign could be found of iron oxide deposits (from reinforcement corrosion) in the wall despite these cracks (Fig 6.5a). In one case, a vertical crack has developed, which appears to be due to an omitted expansion joint (Fig. 6.5b). A number of the stub columns, carrying timber columns supporting the jetty super structure, have cracked at the top. This cracking appears to be directly due to corrosion of the horseshoe steel connection (for the timber columns) which is embedded in the stub columns (Fig. 6.5c). The concrete surface bed around the pool has been stripped of cement paste exposing the aggregates. On the breakwater wall, though there is extensive cracking and concrete delamination near the top edge of the wall, the delamination appears to be due to corrosion of embedded reinforcement (Fig. 6.5d)

(a) Cracked concrete retaining wall

(b) Vertical crack in retainining wall

(c) Stub column cracking

(d) Breakwater having surface concrete material loss, cracking and delamination

Fig. 6-5: Showing various defects at the Langstrand jetty

6.2.1.3. Dolphin Beach Platform

A water park has been built adjacent to Langstrand beach on an area called Dolphin Beach. A timber platform has been erected for an aerial view of the beach area. The platform is carried on a series of reinforced concrete stub columns (Fig. 6-6). An inspection of the structure showed severe delamination and spalling at the top and along the length of the stub columns. The cracking in the top of the concrete could be due to the corrosion seen or the corrosion is a result of the cracking. The timber platform is rigidly fixed to the top of the concrete columns and the platform is braced in one direction. This would imply the top of the columns are under substantial stress from the fixed joint and the platform sway could lead to cracking at the column top. That being said the columns have developed cracks along the column as well. The cracking along the column length suggest there is a stress from within which is leading to the cracking. Concrete pegs around the platform (Fig. 6-7(c)) have developed similar cracks though more extensively. The all reinforce the conclusion that this extensive cracking is due to corrosion of reinforcement bars within the columns. (Fig. 6-7).

Fig. 6-6 – Dolphin Beach platform

(a) (b) (c)

Fig. 6-7: Cracked stub columns (a) & (b) and cracked peg (c)

6.2.2. Bridges

6.2.2.1. Walvis Bay B2 Overpass

The B2 road into Walvis Bay from Swakopmund has an overpass that ensures seamless flow of traffic to the northern part of the town. The overpass is a 3-span concrete beam bridge that straddles over the industrial access road and railway line. The bridge has a mark that shows it was built in 1969. The bridge deck is made up of beam and slab structural system which is supported by abutments on the North and South with two piers in the middle. The bridge's longitudinal beams rests on transverse beams carried by the column piers (Fig. 6-8).

Fig. 6-8: Walvis Bay B2 Overpass Bridge

An inspection of the bridge identified deteriorations that had occurred (Fig. 6-9). The bridge suffered damage from impact on the soffit of some beams which stripped the reinforcement cover from the impacted areas. On the northern end of the bridge there appears to be bearing damage which has led to crashing of the edge beam. The crushing appears to have led to the development of a diagonal crack on the top of the abutment (from the end of the crashed beam). These two defects have so far not shown signs of reinforcement corrosion, but they expose reinforcement to corrosion. The sidewalk and the concrete parapets on the deck have concrete spalling due to reinforcement corrosion, and it is extensive. Similarly, a retaining wall at the southern edge of the bridge footing shows extensive concrete spalling. This retaining wall is to prevent sand from spilling onto the underpass road/rail.

(a) Impact damage to bridge soffit

(b) Abutment with a diagonal crack

(c) Concrete delamination on the sidewalk (d) Retaining wall concrete spalling and delamination

Fig. 6-9: Walvis Bay B2 Overpass Bridge

6.2.2.2. Swakopmund River Bridge

This bridge is built across the Swakop River carrying the B2 highway from Swakopmund to Walvis Bay. The prestressed concrete bridge spans about 600m over the river valley, one of its main features is the 75m main span (Fig. 6-10). Precast concrete girders are simply supported over reinforced concrete pier heads and stiffening reinforced concrete beams span between the girders. An in-situ concrete deck has been cast between the precast T-beams. The bridge does not only offer the vital role of connecting the two municipalities, but it has become a significant tourist landmark for the area, and its durability is essential. The bridge was opened in 1969 according to the inscription on the parapet.

Fig. 6-10: Swakop River Bridge

During the inspection, some concrete deterioration was found (Fig. 6-11). A number of the piers were noted to have developed vertical cracks near the pier foot. It was also observed that all piers had been coated with a cementitious grout and in the area where cracks had come up the grout was peeling off. A number of pier heads had developed crazing on the grouted surface. A second observation made was that a majority of the stiffening beams had developed cracks. However, there was no evidence of corrosion of reinforcement bars.

The parapet of the bridge is made up of a concrete upstand and a steel handrail with its support embedded in the upstand (Fig.6-11b). It was noted that the concrete has severely cracked with spalling evident at almost all steel handrail support points. The origin of these cracks appears to be the embedment which has corroded leading to concrete cracking and corrosion of parapet reinforcement. The coating noted on the piers was also found on the parapet further reinforcing the belief there were repairs done to the piers (Fig.6-11a). Cracking in the grouting at some of the piers shows signs of corrosion damage. The bridge also carries a water supply pipeline, and one manhole of the waterline (on the northern end of the bridge) was noted to have severe concrete delamination due to corrosion (Fig.6-11c).

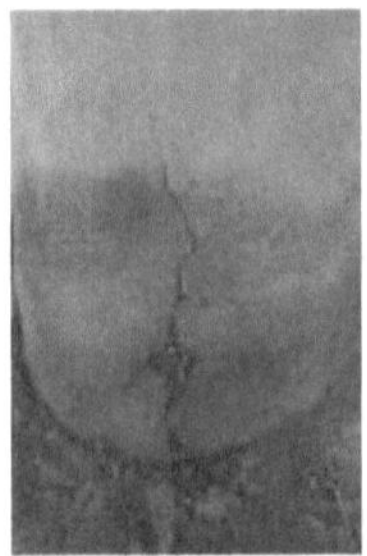

(a) Cracked pier with a grout coating

(b) Cracked parapet upstand beam

(c) Water manhole with concrete delamination

Fig. 6-11: Concrete defects on the Swakop river bridge

6.2.3. Water tanks

Two different water tank sites were visited in Walvis Bay, one located off the C14 road (20 km south-east of town) and another next to B2 near Aphrodite beach (Fig. 6-12). The tank off Aphrodite beach was easily accessible. This tank is partly buried with only a portion of it above ground. An inspection showed the tank to have no sign of any structural distress. The main observation though was the coating that has been applied to the walls. It was also noted that stainless steel pipes were embedded into the tank walls. Use of stainless steel pipe is likely to resist any early corrosion attacks due to its non-reactive nature to corrosive agents.

Fig. 6-12: Aphrodite beach water tank

The tank off C14 is a 15,000 m^3 capacity concrete ground tank. This tank is located within the Namib Desert and away from commercial areas (Fig. 6-13). A full inspection was not possible due to access limitation, but the limited inspection showed that the cover slab of the tank has cracking and spalling due to reinforcement corrosion. Though the tank is within a desert environment its location is within the Atlantic fog effect zone highlighted in 2.2. At mid-height of some tank sides leaks were noted (Fig. 13). The mid-height leakage were seen coming from horizontal cracks which appears to possibly be cold joint in the concrete.

(a) Off C14 Water tank

(b) mid-height leakage

(c) Corrosion damage on tank roof slab

Fig. 6-13: C14 Tank and concrete damage to tank

6.3. Swakopmund

6.3.1. Water tanks

Three different water tank sites were inspected in Swakopmund. The first site is located East of the town just off of the B2 road. This site has an old 20,000 m^3 rectangular tank and a newly built circular tank; both of these are ground tanks (Fig. 6-14). The old tank was noted to have severe reinforcement corrosion-related damage. The corrosion of the reinforcement has led to spalling of concrete on sections of the tank. This spalling might have led to the decision to build the new circular tank adjacent to it. The leakages observe on this tank appear to come through vertical joints in the tank wall which suggest a failure of water stops used in these joints.

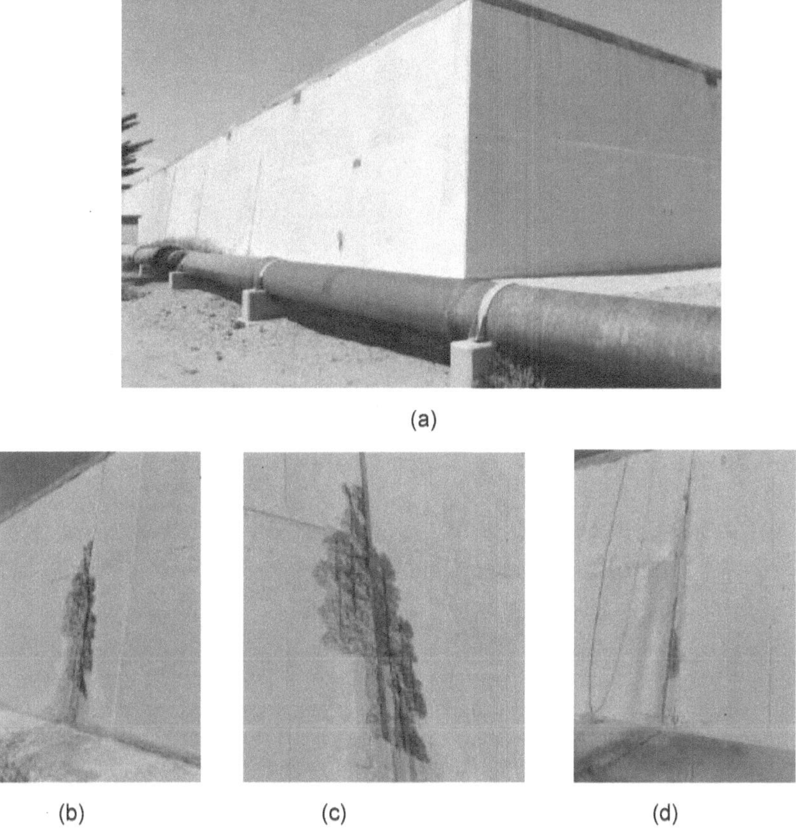

(a)

(b) (c) (d)

Fig. 6-14: Swakopmund B2 20,000 m^3 tank concrete defects

The second tank site is located near the Swakopmund airport East of Mondesa. The site was partially accessible, but even with the partial access, the concrete defects on the tank were visible. It could be seen that there has been corrosion of reinforcement on localised spots on the wall of the tank with spalling at the spots (Fig. 6-15). The position of the corrosion is on horizontal joint position which points to possible joint failure. There is also concrete spalling at the foot of the tank due to reinforcement corrosion. The area at the foot of the tank is generally damp from the early morning coastal humidity. The dampness would easily exploits any defective concrete surfaces to initiate and accelerate reinforcement corrosion.

Fig. 6-15: showing concrete defects on Mondesa tank

6.3.2. Swakopmund Wastewater Treatment Works

The old Swakopmund sewerage treatment works was opened in 1953 and has seen several upgrades until 2000 (Van Dyk, 2012). The works is made up of pump stations, activated sludge beds, drying beds and trickling filters. Most of these are reinforced concrete structures except for the pump house and offices (Fig. 6-16). At the end of the last upgrades, a decision was made to build a new treatment works to complement this old facility. According to Van Dyk (2012), the main reason was town expansion that had overrun the facility making expansion impossible, on completion of the new facility the two are being used to handle wastewater for the town.

Fig. 6-16: Aerial view of Old Swakopmund Sewerage treatment works (Google map 2020)

An inspection of the site showed that almost all reinforced concrete structures have severe corrosion-related deterioration except the newly built pump house. Several facilities have also been abandoned (Fig. 6-17). Corrosion damage was noted both on the outside of the structures as well as inside of some. Sloping roofs for the digesters were found to have severely cracked surfaces though delamination had not started yet. The crazed cracking identified on the digester roof (Fig. 6-17(d)) appear to be due to concrete shrinkage. There are also cracks along horizontal joint positions on the digester roofs. The waste treatment ponds on the other hand show signs of corrosion damage to the concrete. The surface of the concrete (Fig. 6-17(b) and (d)) shows signs of cement paste loss in areas where signs of reinforcement corrosion is present. This is consistent with biogenic attack prevalent in sewerage treatment works (Ng P.L., Kwan A.K.H., 2015). There is a difference in the amount of cement paste loss and corrosion consequence between the bridge over the pond (that has more paste loss) and the external surface (that has less loss but more spalling of concrete). This could be explained due to high exposure to the fact the bridge surface is directly within the headspace of any hydrogen sulphide that would be escaping from the pond. Thus the bridge surface is under constant cement stripping conditions, meanwhile, the pond external surface is affect by settling has to also contend with high chloride containing groundwater.

(a) cracked pond with wastewater

(b) cracked empty water treatment pond

(c) Cracked digester roof

(d) Pond bridge with corrosion damage

(e) Cracked pump house roof

(f) Newly built lift station

Fig. 6-17: Concrete state in the Swakopmund Sewerage Treatment works

6.3.3. Swakopmund B2 Road over rail bridge

At the eastern edge of Swakopmund is a bridge overpass built to span over the main railway line. The bridge is a 2-span concrete beam-slab bridge. During an inspection, it was found that the concrete deterioration had rendered the bridge unstable that it required temporary supporting (Fig. 6-18). There are five (5) pier columns and the outer ones had vertical cracks running through their height. Spalling of concrete and delamination of reinforcement were found on all columns. The edge beams had horizontal cracks running the length of the beam. The abutment walls have severe vertical and horizontal cracks with concrete spalling in several places. A week after this inspection was undertaken, the road was closed, and the bridge was demolished with a new one to be constructed (RA, 2019). The cracks in both the column and the beams followed the line of the main reinforcement in the elements which suggests corrosion induced cracking. The bridge is in an area with high salt-laden humid condition implying this corrosion is most likely chloride-induced corrosion.

(a) Falsework erected under bridge

(b) Cracked bridge beam

(c) Cracked abutment wall

(d) Cracked wingwall

(e) & (f) Temporary vertical support

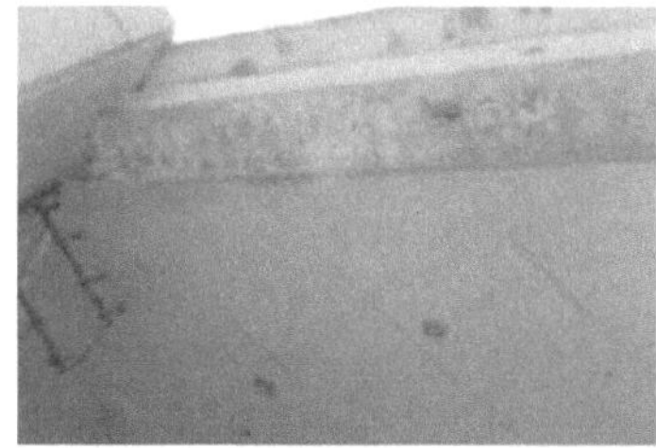

(g) Cracked beam

Fig. 6-18: – Swakopmund B2 Overpass and concrete deterioration

6.3.4. New Swakopmund –Walvis Bay behind Dune 7 road

The road between Swakopmund and Walvis Bay behind Dune 7 (which is within 10Km from the ocean) is currently being upgraded, and part of the project is the construction of several bridge overpasses. Fig. 6-19 shows selected pictures taken from some of the overpass works. What is of interest is the exposure of the site to corrosive foggy environment which can be seen engulfing the overpass under construction. In Fig. 6-19(a), the parapet starter bars are left exposed to the foggy environment. In Fig. 6-19(b), the pier, abutment and decking reinforcement can be seen exposed to the weather elements. The exposure of reinforcement has two potential problems which both could affect the service life of the bridge structures. First chloride the fog that settles on the reinforcement bars could lead to development of half-cell potential areas on the rebar. Secondly, the fog as it settles on the cast concrete, there is a potential to raise the chlorides in pore solution within the concrete dependent on how long and how much fog has fallen when concrete is still plastic. Both these can lead to earlier than anticipated loss of service life (as presented in 3.8.1) and highlights part of the challenge of concreting in these coastal area.

An experimental study by Bagheri et al (2019) investigated the effect of early age ingress of chlorine into concrete (results of which could be related to salt-laden fog settling on newly cast concrete amongst others). The study showed that concrete exposed to chloride at early age (from immediately after casting) suffers a loss in service life that is higher than previously thought due to increase in surface chloride value. Though this was a study on a small scale but it points to an issue that has received little attention.

(a) C28 overpass on a foggy morning

(b) Swakop river bridge piers and abutment

Fig. 6-19: Bridges on the Swakopmund – Walvis Bay road construction project

6.4. Usakos

6.4.1. Water tanks

Two water tank sites were inspected in Usakos. The main service tank and a smaller tank off the B2 road at the eastern end of town. The smaller tank was found to have a cementitious coating on the surface, but no reason was seen for this application. It was noted that the walls appeared to be in good condition though the roof had concrete spalling near the overflow pipe of the tank (Fig. 6-20). The roof slab detail was found to have a groove on the soffit of the protruding edge, which was not seen on any of the tanks inspected. The spalling of the concrete appears to have originated from the area of the groove.

(a) Off-B2 water tank

(b) Cracked roof slab edge

Fig. 6-20: Usakos off B2 road water tank

The main supply tanks for the town are located on a hill north of the town. The site was partially accessible, but it could be seen that two tanks were on the site with an old (1150 m^3) and a new one (3000 m^3). The old tank was noted to have a cementitious coating around it, but horizontal cracks were still visible with leakage around mid-height one (Fig. 6-21). The horizontal cracks suggest cold joint development at these construction joints positions and possibly failure of the water bars.

(a) Main supply tanks on the hill overlooking Usakos Town

(b) New main tank

(c) Old main tank

Fig. 6-21: Usakos town main tanks

6.4.2. Road Bridges

Two road bridges were inspected in Usakos, one on the eastern-end and another on the western-end of the town along the B2 road. The eastern end bridge is a 3-span concrete beam bridge with abutments on either end and pier walls supporting the deck. An inscription on the parapet shows that the bridge was opened in 1966 (Fig. 6-22). An inspection revealed that the bridge is mostly in good condition though some defects were found. The worst defects were found on the northern abutment which had a vertical crack at midspan. There was also some horizontal plastering, on the abutment, that appears to be more recent works. The wingwall of the same abutment had some plastering work which had cracks. One of the pier walls was found to have a honeycombed section near the bottom which appears to be part of the original works. It was also noted that the piers and pile caps on the eastern face had surface material stripping.

Adjacent to the bridge is an old gravel road that was used before the construction of the bridge. On this road, there is a concrete pipe culvert headwall which is heavily cracked with reinforcement exposed at a few places though no delamination/spalling (of the concrete) or severe corrosion of the reinforcement was noted (Fig. 6-22e). The cracks appear to be shrinkage cracks from the time of construction.

(a) Usakos East B2 bridge

(b) Vertical crack on abutment wall

(b) Patched wingwall with cracks

(c) Honeycombed pier wall

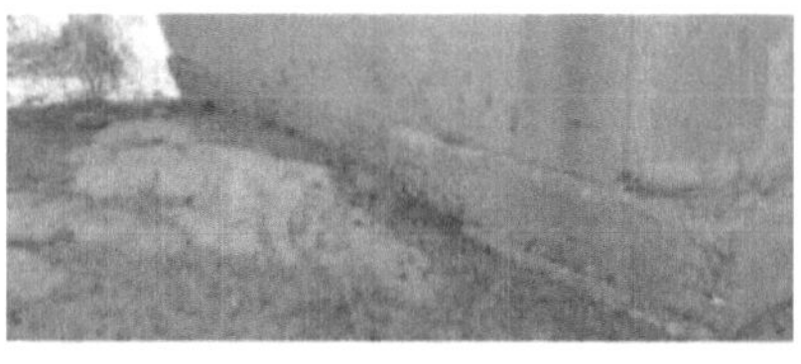

(d) Pile cap stripped of surface material

(e) Cracked headwall

Fig. 6-22: Usakos East B2 Bridge and concrete deteriorations

The western bridge is a girder bridge about 130 m long with a total of eight spans. The bridge was opened in 1952, according to an inscription on the bridge. Inspection of the bridge showed that it was in good condition except for some cracking on the southern abutment wingwall. There also was stripping of surface materials on the middle four pier walls (Fig. 6-23). One feature noticed on this bridge was an expansion joint that was introduced along the bridge centreline through the deck and all vertical elements.

Fig. 6-23: Usakos west B2 bridge.

6.4.3. Rail over road bridge

South-east of Usakos town there runs a railway line (between Windhoek and Walvis Bay port) which has been built over a main road in the town, D1914. The railway is carried by a composite concrete steel 2-span slab bridge (Fig. 6-24). The vertical elements are all reinforced concrete elements, whereas the deck is made of a concrete slab that is carried on steel beams. Inspection of the bridge showed several defects. The deck is supported over the pier, and it was noted (when a train passed by) that the deck vibrated quite substantially bouncing on the bearings. This possibly would explain the concrete crushing noted at midspan of the deck on either bridge face. It was noted that there were horizontal cracks that have radiated along the deck towards the abutment from the crushed area. The wingwalls were found to have diagonal cracks radiating from the top near to the point of the deck support.

(a) Usakos Rail over road bridge with chipped parapet foot

(b) Deck damaged

(c) Wingwall with diagonal cracks

Fig. 6-24: Usakos Rail over Road bridge

6.4.4. Abandoned brick factory

East of Usakos, there is an abandoned factory (Fig. 6-25a) used for making bricks and precast concrete elements. Locals mentioned that the site was abandoned at least five years ago. There are several concrete structures amongst these ruins, and they point to the corrosion rate within the town. The concrete structures appear to be made up of composite steel construction. It was noted that in cases where steel has been exposed the surface is covered by corrosion by-products but there is no sign of material loss or delamination (Fig. 6-25b). There were also no signs of deterioration of any concrete elements due to corrosion, even in areas where coverage could be seen to be minimal.

(a) Abandonded factory building

(b) Exposed steel embedded in concrete

(c) Embedded steel with no concrete cover

Fig. 6-25: Abandonded Usakos building with exposed structural steel

6.5. Arandis Bridges

Arandis town is located 50 km East of the Atlantic coast. Three bridges, located just outside of the town, were inspected. The first bridge looked at was a 3-span slab bridge at the entrance to the town that is built over the railway line (Fig. 6-26a). On this bridge, the main deterioration noted was on the northern end of the bridge. It was observed that there was a concrete crushing failure at the abutment wall support, which seems related to the bearing collapse at the abutment. The pier next to the abutment has a section of concrete delamination at the foot (Fig. 6-26C). One other interesting observation was a marker that had concrete stripped from the reinforcement that stood between the rail tracks. While the concrete showed signs of ageing, the reinforcement had no serious corrosion problem (Fig. 6-26b).

(a) Arandis Road Bridge over rail

(b) Cracked rail marker

(c) Crashed abutment

(d) Cracked pier

Fig. 6-26: Arandis Road over rail and concrete defects

The second bridge inspected was the B2 highway bridge over rail; it is a single-span concrete slab bridge (Fig. 6-27a). The inspection showed cracking in the abutment wall and wing wall. There appears to have been repairs done to the cracks using an epoxy-based material. No visible evidence of corrosion of the reinforcement were noted in the cracked areas (Fig.6-27b and c).

(a) Arandis Road over Rail bridge

(b) Abutment wall with repaired cracks (c) Wingwall with a diagnoal crack

Fig. 6-27: Arandis B2 Road over Rail Bridge and concrete deterioration

The third bridge inspected was the B2 bridge built over the Rossing industrial road and rail. This bridge is a concrete slab bridge with 3-spans (Fig. 6-28a). The inspection found that the pier had a crack at the top and a cementitious grout had been applied on it. There was also a smoke-stained soffit of the deck over the rail portion of the bridge (Fig. 6-28c). A grid marking was found on this soffit, indicating an investigation was done to assess the cover to the reinforcement (Fig. 6-28b). There are probably two reasons for this smoke staining; either there was a fire or trains use this as a shunting spot. The white patch next to the black stained surface suggests increased heat on the concrete soffit, which could be a product of both suspected causes.

(a) B2 Bridge over Rossing Road and Rail

(b) smoke stained deck soffit

(c) Cracked pier top

Fig. 6-28: The Rossing B2 Bridge over rail and road

6.6. Discussion

The visual inspections on the different structures (in different places) reaffirmed that over time concrete structures would develop defects. The various structures had a wide variety of deterioration conditions. Several conclusions can be drawn from the observed concrete conditions. With the known theories of concrete deterioration (Chapter 3), it is feasible to theorise on the causes of the deteriorations observed and the varying rate of deterioration.

6.6.1. Functionality

The inspection showed that deterioration of concrete occurred in concrete structures with different functionalities. In general, water containing structures and bridges were inspected, and all exhibited different levels of deterioration. Almost all the severe deteriorations observed were directly linked to corrosion of steel in concrete. Corrosion damage with different levels of concrete spalling were prevalent in the different structures.

6.6.2. Location

The inspection results re-enforce the corrosion damage risk that structures built along the coastal town are under. Virtually all structures inspected in the two coastal towns exhibited substantial corrosion-related damage. Chapter 2.2 highlighted the humid (sea-water filled) conditions that are prevalent in the two towns in contrast to the desert towns of Arandis and Usakos.

The worst deteriorated structures were found in Swakopmund. This was a little surprising; it was expected the two coastal towns would have comparable deterioration as they enjoy virtually similar environmental conditions. Also, age of the structures in the two towns are virtually the same, but corrosion appear to have affected Swakopmund much more severly than Walvis Bay. The research could not identify why this was so.

6.6.3. Design related defects

Some of the defects seen on site could be classified as design defects

a) Expansion joint location (or lack of) – vertical cracks seen at Langstrand jetty and some bridge abutment walls could be attributed to missing expansion joints. In the case of the Langstrand jetty, it was easy to see as other areas had vertical expansion joints equally spaced. The area without an expansion joint ended up developing a vertical crack within the area the joint should have been. For the bridge abutment it was noted in Usakos that the main bridge had an expansion joint along the deck which went into the pier and abutment wall and no vertical crack developed in the wall. The eastern Usakos bridge (and other bridges) without an expansion joint developed a crack in the middle of the wall, and thus the crack could be traced back to lack of expansion joint.

b) Bearing defect – in the cases where bearing failures were noted, the author questions whether the bearings used were the right ones. Shiua et al (2008) noted that bearing design defects can include inappropriate bearing choice, inadequate bearing space, poor material choice and lack of corrosion protection. These design errors get exposed with crushing failure during the service life of a bridge. However, the ultimate effect of crushing is development of cracks which exposes reinforcement to corrosive environment.

c) Embedments – bridges and jetty structures in the coastal region that had metal embedments suffered substantial corrosion-related cracking around the embedments. The observed defects indicate that the metal embedment chosen were not resistant enough to corrosion. This led to initial cracking of the concrete that exposed embedded reinforcement leading to their corrosion followed by corrosion propagation and delamination/spalling of the concrete cracks.

d) The impact damage that concrete structures suffered suggested design deficiency in choice of clearance/protection for bridges and small craft jetty. Bridges that were found to have impact damage on the soffit of the beams were all in industrial areas which leads to questioning of the provided clearance. It has to be acknowledged that since most of the bridges were built in the early 1970s, there is a possibility that changing traffic regulations might have affected loads carried by vehicles. That being said when ergulations have been revised it is expected that a review of impact of the change is undertaken and put in place necessary mitigations to prevent damage of these structures. The small craft jetty had edges exposed to vessel impact due to the use of improvised fenders.

e) Environmental exposure consideration – though no defect was observed that could directly be related to this, as presented in 6.3.4, the specification should have considered the environmental conditions that might be prevalent during construction of new structures. This failure is a potential to reduced service life.

6.6.4. Construction defects

Some deteriorations observed are due to the construction errors

a) Development of cold joint – a number of the structures were found to have developed horizontal joints which the author believes were cold joints. In the case of water tanks inspected, the old tank for Swakopmund, Walvis Bay and Usakos leakages were seen from these joints. These leaks provide water that is a critical ingredient in the corrosion of reinforcement. Similar joints were seen on the jetty at Langstrand.
b) Minimal cover to reinforcement – whereas during the investigation no full cover measurements were done for the structural elements, visually it was noted that in several structures where concrete had cracked, and reinforcement was exposed, the amount of cover provided had been minimal. This was seen across all structures where the reinforcements were exposed.
c) At the old Swakopmund, sewerage treatment works, and the jetty at Langstrand concrete shrinkage cracks were prevalent on the concrete faces. Whereas at the treatment works these were prominent in slope faces for the pool wall at the Langstrand jetty these were on a vertical structure.

6.6.5. Maintenance failures

Several structures were found to have had grout applied on the cast concrete surface, and there was evidence that this material was applied as part of a maintenance protocol on defects that had developed. It was, however, found that the underlying issue was not sorted. For instance, the Swakopmund bridges spalling of concrete continued after the grout had been applied; in the case of the road over Rail Bridge spalling became so severe leading to the bridge being demolished. The failure of the maintenance to arrest the underlying corrosion problem is thus seen as the cause of the current defects observed.

Defects seen in some of the water tanks can be attributed to operational maintenance defects. In Usakos and Swakopmund, there was consistent leaking of water from the water inlet and overflow pipelines. In Swakopmund, tanks were found to have leaking expansion joints. The reinforcement corrosion and concrete spalling on these tanks were a direct result of the humid conditions created on the concrete surface.

7. Concrete testing in Namibia

7.1. Introduction

Testing of concrete and concrete constituent materials in Namibia is undertaken either by commercially available material laboratories or material laboratories established by contracting companies (as part of a project implementation team). The universities of Namibia and Namibia University of Science and Technology have material laboratories that are not available for private material testing. Similarly, the cement producers' material laboratories are not open to testing of concrete works other than in-house needs. As part of the research, interviews were conducted to assess capability, challenges and adaptability to concrete durability testing. Interviews were undertaken with civil engineers to assess awareness of durability and performance-based concrete design and specification. In November 2018, Ohorongo Cement organised a day seminar discussing various concrete-related issues, including durability and performance-based durability design. Observations from this session have been included in the knowledge assessment.

7.2. Commercial material laboratories

A total of four commercial laboratories (which are undertaking concrete testing across the country) were interviewed. The laboratories are shown in Table 7-1.

Table 7-1: Commercial material laboratories interviewed

Item	Laboratory	Location	Date interviewed
1	Access laboratory	Windhoek/Swakopmund	12/06/2019
2	Namibia Technical Services	Windhoek	14/06/2019
3	ADCON	Windhoek/Swakopmund	12/06/2019
4	Namibia Civil Engineering laboratory	Windhoek	14/06/2019

Material testing services provided by these laboratories are offered because of the commercial sense of test. It was found that all these laboratories principally offer only concrete slump and compression strength testing service. One laboratory, ADCON, was said to have offered concrete cover measurement and potential mapping service. Interviews with all laboratories showed there was a lack of familiarity with the South African DI method.

7.3. Contractor established material laboratories

Six major civil construction projects that were running (in different parts of Namibia) were looked at, and Table 7-2 presents these projects and the main concrete components. Based on discussions with project engineers and also direct observations, a conclusion was made on the material testing practices. It was found that these laboratories are specially established to provide a readily available material testing service to comply with project requirements, i.e. quality assurance. The laboratory equipment present is thus specifically for those tests the contractor is asked to provide by the project specification being implemented. Of all these projects, only the NAMPORT projects called for durability testing.

Table 7-2: Major Construction projects reviewed

Item	Project name	Agency	Location	Main concrete features	Date interviewed
1	Port of Walvis Bay – New container terminal	NAMPORT	Walvis Bay	Marine structures including in-situ bored piles, precast piles and quay structures	27/03/2019 12/04/2019
2	National oil storage facility – Walvis Bay project	NAMPORT	Walvis Bay	Jetty structure with in-situ bored piles and RC superstructure	12/06/2019
3	Construction of Neckartal Dam	Ministry of Agriculture	Keetmans hop	Roller compacted concrete dam structure	17/05/2019
4	Windhoek – Hosea Kutako International Airport to freeway	Roads Authority	Windhoek	Bridge structures	03/04/2019
5	Upgrading to Dual Carriage-Freeway standards of TR 1/6: Windhoek – Okahandja	Roads Authority	Okahandja	Bridge structures	03/04/2019
6	Swakopmund Reservoir construction project	NAMWATER	Swakopmu nd	20,000 m^3 Water tank constructio n project	17/05/2019

China Harbour Engineering Company (CHEC) is implementing two (2) projects using the NAMPORT concrete specification. As presented in 4.3.3, these specifications require the undertaking of durability related tests, including South African DI tests. CHEC confirmed that for the DI tests, samples had to be sent to UCT's COMSIRU laboratory for testing as the CHEC laboratory did not have the appropriate test equipment and engineers were not very familiar with the test methodology. At the same time, CHEC laboratory confirmed that the laboratory was able to undertake some durability tests called for in line with ASTM.

7.4. Materials engineering practitioners

Several conclusions were made from discussions with and observations of materials engineering practitioners

a) Engineers interviewed acknowledged the importance of curing for concrete durability. It is generally acknowledged that curing is one of the most ignored construction activities unless a strict supervision strategy is implemented on construction sites.
b) Virtually all engineers interviewed lamented the lack of an assessment tool to confirm that curing has been appropriately undertaken on site.
c) For engineers working in coastal areas, there is an acknowledgement of the high corrosiveness of the area though none could provide technical advice notes as reference on working in these areas.
d) Concrete deterioration is a subject that is not fully understood though several engineers have taken post-graduate courses in it and have gone to make changes where they are working. A good example is at NAMPORT where after this (concrete deterioration and performance specification) knowledge acquisition the port engineer's office pushed for the revision of the agency's concrete specification to cover more durability matters than before.
e) The Namibian construction industry has not provided enough opportunities for studying concrete deterioration and durability locally. The few people that have acquired further knowledge have had to travel outside for the training.
f) There is a heavy reliance on the use of coatings in making concrete less permeable to water. However, the coating does not change the permeability of concrete only offering a temporary barrier against ingress of aggressive agents if applied correctly. Correct application of a coating involves correct choice of coating material, preparation of the concrete to be repaired, correct mix of the coating, applying a proper thickness of the coating and curing of the applied coating. In the absence of local analytical guidance on this, engineers are reliant on advice from material supplier technical specifications which in many instances is shared by salespeople with a bias to selling a product.

7.5. Conclusion

The research showed that there are a number of institutions undertaking material testing in Namibia. The testing capabilities being offered by these laboratories are limited principally because of the tests demanded by the market which are in line with the prescriptive approach. In two project where durability index testing was required samples had to be sent to South Africa for testing which is not sustainable if such testing regimes are demanded across all projects. There is need of training in deterioration mechanisms, concrete repair and performance based durability design and construction specification. There is generally a willingness to acquire new knowledge in the industry.

8. Concrete infrastructure state analysis

8.1. Introduction

Concrete defects identified on the inspected structures (in the four towns) can be directly linked to the environment that the structures are built in. It can also be shown that deficiency in the design and construction specification lead to some of the problems noted.

8.2. State of concrete infrastructure and the environment

The inspected sites can be split into two groups; coastal towns (Swakopmund and Walvis Bay) and desert towns (Arandis and Usakos). The predominant defects identified followed this distinct split. Of the defects, identified reinforcement corrosion was found to have caused the most damage to the structures.

Structures in the coastal town virtually all had signs of corrosion damage to varying degrees. The coastal areas are humid places with little or no rain. Chapter 2.2 highlighted that these areas receive a lot of fog (that is saturated with salt solution) which makes the areas highly corrosive to steel. Reinforcement should receive protection from the concrete it is embedded in. However, when the fog settles on the concrete surface, salt solution (NaCl) is deposited on it. Multiple transport processes enable the ingress of the solution into the concrete. The corrosion damage shows that the chloride ions from the solution does reach the reinforcement level leading to deterioration. The environment, in this case, continuoslly supplies salt solution which provides both the water and the chloride ions for the corrosion.

The water tank in Usakos had corrosion damage on the roof edges (Chapter 6.4.1). Though considered an anomaly, this damage highlights the risk that concrete structures have even in such an arid environment. The dripping water created a damp environment that enabled water to be available for the corrosion of the reinforcement. The depassivation of the concrete could have been due to a number of reasons. The potable water (dripping from the tank) contains chlorine which might have led to the depassivation. There is also carbon available for carbonation.

8.3. State of concrete and concrete design/specification

Following the recommended design and construction specifications is supposed to produce concrete that will minimise permeability of concrete and maximise resistance to aggressive agents (Ballim et al 2009). The corrosion defects identified highlight the fact that different areas will have different durability demands on concrete. Thus failure of having unique performance requirements makes concrete structure vulnerable to durability failure (in this case reinforcement corrosion). In drawing this conclusion the author appreciates that the investigations were carried on structures which at the time had stood for over 25 years. However, the repairs seen on the structures in Walvis Bay and Swakopmund suggest these durability failures occurred long before the investigation. Unfortunately the bulk of the repairs were improperly undertaken.

With all coastal structures suffering corrosion damage, it shows all concrete to a degree allowed for ingress of chloride ions and water to reinforcement level leading to corrosion. Chapter 4.3 highlighted specifications that are followed in the construction of most concrete structures in this area. A commonality of these (specifications) is that strength of concrete is an acceptance criterion and concrete mix design has limitations (maximum water/binder ratio and minimum cement content). Practitioners have believed that increase in binder content would produce higher strength (and ultimately more durable) concrete. This is something Alexander et al (2017) reinforced as being incorrect. One would postulate that these coastal structures emphasised on achieving strength requirement (as specified) at the detriment (unknowingly) of producing concrete that was impenetrable to salty water that it would be exposed to daily. Alexander et al (2017) showed that there is limited benefits to durability with an increase in binder. The minimum binder content, in the specification, if followed has the potential of leading to production of permeable concrete which leaves concrete at higher risk to reinforcement corrosion.

The specifications currently do not demand confirmation that the concrete produced will be durable. For practitioners, then the only option is to wait and see how the concrete performs in practice. The structures inspected suggest concrete produced is highly permeable that in areas with corrosive agents, the reinforcement gets affected.

8.4. State of concrete infrastructure and practitioners

Assessing some interventions undertaken in the management and operation of infrastructure provides a glimpse into capabilities of practitioners. The whole essence of maintenance/rehabilitation is to improve performance of a structure and extend its useful life, as shown in Fig. 5-1. Several issues were learnt from this

a) The small craft jetty was the only structure that a proper condition assessment was done followed by a systematic maintenance plan. External international expertise for measurement of the depth of chlorination was utilised. This tied in with findings (Chapter 7) that showed that local material testing agencies do not possess the capability to undertake durability tests on cast concrete.
b) Swakop River Bridge and Swakopmund B2 bridge-over-rail had received corrosion damage repair works. In both cases, the repairs appear to have been superficial. On the B2 Bridge maintenance works had just recently been undertaken when the decision to bring it down was made whereas the Swakop River bridge cracks have appeared above repair works. The author theorises that these repairs point to knowledge gaps in the cause of the corrosion defect and how to repair it.
c) The Swakopmund wastewater treatment (structures) corrosion damage was found to be more on the outside of the concrete structures; the few internal surfaces inspected had isolated visible corrosion damage. It would be necessary for designers to appreciate the different environments the external and internal

surfaces (of these structures) get exposed to. Whereas biogenic corrosion would be prevalent on the 'inside' of the elements (including those striding over these structures) external surfaces in contact with the ground have to contend with biogenic effect, fog and ground water.

d) The Wastewater treatment works has several concrete structures in a state of severe delamination and spalling, and no repair works seem to have been attempted on them. In the current state, the structures are enroute to being demolished. Fortunately, the town has an alternative facility but as this (old treatment works) is still operational, its collapse will have disruption to the municipal sewer system. The B2 Bridge over rail (when it was demolished) caused a year's disruption to transport flow into and out of Swakopmund town. This exhibits the disruptive effect of lack of or inadequate maintenance/rehabilitation of corrosion damage to concrete structures.

9. Conclusion

9.1. Introduction

Concrete structures play an essential role in the Namibian infrastructure development fabric, and their durability is vital to the national economy. The durability of concrete has been shown to be dependent on the interaction of the concrete with the environment. This interaction is influenced by the design and construction principles/techniques followed. This research looked at standards (for concrete durability design and construction) and how existing concrete structures have performed in Namibia.

Concrete structures in selected four towns were inspected and their conditions assessed (and causes of deteriorations identified). The concrete structures inspected were built a long time ago and most likely not using the current design standards and construction specifications being used in Namibia. Despite this, the deteriorations observed highlights risks that concrete structures built in Namibia have to perform under.

In the absence of a national standard for design and construction supervision, the research has shown that development agencies use different standards to undertake concrete works. NAMPORT, RA and NAMWATER are amongst leading concrete structure developers in Namibia, and the research has shown that all have been using prescriptive methods. Recently NAMPORT has adopted performance-based specifications which have only been used on two on-going projects. The review highlighted the capabilities and inadequacies of current practices.

The research further assessed the knowledge and capability of practitioners. It specifically looked at knowledge and capability with regards to concrete durability. Based on this, limitations and shortfalls were noted.

With the knowledge, from the research, several conclusions can be drawn and recommendations can be made to improve the situation. This (conclusions and recommendations) would assist in ensuring more durable concrete structures are built and continue aiding in the country's economic development.

9.2. Conclusions

9.2.1. The prevailing concrete durability risk

The concrete structures inspected exhibited varied deteriorations; however, the most severe were related to reinforcement corrosion. Coastal structures exhibited the most significant reinforcement damage, whereas one of the inland structures had this problem. For the coastal structures, the corrosion of the reinforcement would directly be due to the prevailing high salty humid conditions (and almost no rain) in the region. However, reinforcement is embedded in concrete. This implies the chloride (from the humid environment) has been able to reach the reinforcement from the concrete surface. For the inland structure with reinforcement corrosion problem, the concrete surface had been continuously made damp due to failure to manage to overflow from the tank. The corrosion suggests the water was able to migrate to reinforcement level leading to the deterioration observed.

This problem though (of ingress of chloride or water or carbonated water) currently is not addressed by the prescriptive design and construction specification being used in construction of concrete structures (and most likely was not addressed in specifications and standards used at time of constructing inspected structures). The research has shown the following

i. The limitation on the concrete mix design takes away the ability of concrete producers to produce concrete that suit specific environmental performance
ii. Concrete strength is the main acceptance criterion for concrete works. However, the ability of chloride to migrate into a concrete structure has been shown to be the most detrimental action on concrete durability. There currently is no requirement for a measure on ability/potential of chloride from ingressing into a concrete
iii. NAMPORT's recently implemented performance-based works specifications have not yet been reviewed to assess effectiveness and difficulties. A review of the implementation would provide insightful lessons could greats help the Namibian concrete industry if they can be shared.

The research thus concludes that the most significant risk to concrete durability is deterioration due to reinforcement corrosion and unfortunately, this risk is currently not adequately mitigated against. This risk is highest along the coastal areas but still prevalent inland.

9.2.2. International durability design and construction advancements

Prescriptive concrete durability specifications have been shown internationally of not always leading to production of durable concrete (Chapter 4). Various earlier than anticipated deterioration led to studies which highlighted shortfalls with the prescriptive approach. The studies have led to a better understanding of the relationship between concrete durability and the environment. This research highlighted advances in performance-based durability specification. This approach confirms performance of concrete using measurable performance requirements.

In South Africa, one country that has advanced substantially in this area, a DI method has been developed for the concrete specification. The method requires testing of specific concrete characteristics that relate to ingress and movement of aggressive agents in concrete. The tests have undergone rigorous testing, including comparison with other international tests to verify the effectiveness and the results have seen the tests being accepted as a South African standard of testing concrete. SANRAL has at present adopted the DI method as a criterion for acceptance of concrete in addition to strength.

9.2.3. Industry and industry standards

Namibia does not have a unified industry standard for design and construction of concrete structures. Construction industry has adopted international standards for use in most cases. In other cases, infrastructure developing agencies have developed in-house standards which are followed. NAMWATER and RA are the main infrastructure developing agencies; constructing and managing concrete structures across the country. Both agencies use construction specifications adopted from South Africa, SANS 1200 and COLTO 1998. These standards use prescriptive specification for concrete construction. The laboratories in the country fully support the testing requirements under these specifications.

NAMPORT has the mandate to develop and manage public concrete structures (for maritime use) across the Namibian coastline. The research found that unlike other agencies, NAMPORT has developed an in-house specification for concrete construction. These are performance-based with requirements for concrete performance specifically against deterioration from corrosion damage. These specifications were developed with the adoption of South Africa's DI method with specific tests required to be performed to show the performance of concrete. Currently the specifications require performance setting at mix design stage with no requirement for compliance check during casting or verification after concrete construction. At present, only two projects have been implementing these specifications, and the effectiveness has not yet been documented. One main challenge faced though has been the capacity of local laboratories to implement tests required under the DI method. The two projects have thus had to export concrete specimens to South Africa for tests.

The research has also shown that practitioners in the industry lack solid understanding of concrete durability and deterioration mechanisms. As concrete durability performance is currently not demanded by the bulk of concrete projects, practical experience is lacking. In addition to this, minimal professional advancement programs are available locally.

References

Alexander M.G., (2016) - *Performance-Based concrete durability design and specification in South Africa-Background, Implementation & Quo nunc?*, FIB Symposium key note address, page 52-62, Cape Town.

Alexander, M.G., Angelucci, M, Beushausen, H., J.R. Mackechnie (2017) - Specifying Cement content for concrete durability why less is more, page 12-16, Concrete Beton Number 50, South Africa

Alexander, M.G., Ballim, Y, and Beushausen, H. (2009) - *Durability of concrete*, Fulton's concrete technology page 201 •, Cement & Concrete Institute

Alexander, M.G., Ballim, Y., and Stanish, K. (2006) - *Assessing the repeatability and reproducibility values of South African durability index tests*, Pages 10–17, Journal of the South African Institution of civil engineering Vol 48 No 2

Alexander, M.G., Basheer, M., Baroghel-Bouny, V., Beushausen, H., d'Andrea, R., Gulikers, J., Gonçalves, A., Jacobs, F., Khrapko, M., Monteiro, A.V, Nanukuttan, S.V., Otieno, M., Polder, R. (2015) - *Principles of the Performance-Based Approach for Concrete Durability*, State-of-the-Art Report RILEM TC 230-PSC

Alexander, M.G, and Baushausen, H. (2008) - The South African Durability index tests in an international comparison, page 25-31, Journal of the South African Institution of Civil Engineer, Vol. 50 No. 1.

Alexander, M.G., Beushausen, H., and Salvoldi, B., (2015), 'The correlation between oxygen permeability and the carbonation of concrete', Construction and Building Materials, Vol. 85 Page 30-37

Alexander, M.G., Baushausen, H, and Torrent, R. (2008) - *Performance-based specification and control of durability of reinforced concrete structures*, page 319 – 323, SACoMaTiS International Conference

Alexander. M.G. (2017), *Durability Index Testing Procedure Manual 2017*, Version 4, University of Cape Town-CoMSIRU, Cape Town

Almaral-Sánchez J.L, Almeraya Calderón F., Arredondo-Rea S.P., Castorena-González J.H., Corral-Higuera R., Gómez-Soberón J.M., Neri-Flores M.A., (2011) - *Sulfate Attack and Reinforcement Corrosion in Concrete with Recycled Concrete Aggregates and Supplementary Cementing Materials*, page 613-621, International Journal of electrochemical science, Volume 6

AQUAFACT International Services Ltd (2012), *Sanitary Survey Report for the Namibian Shellfish Sanitattion Monitoring Programme*, page 6-8, The Namibian Standards Institution, Windhoek

Beeby, A.W., Forrest, J.C.M, Harrison, T.A., Menzies, J.B., Moore, J.F.A, Newman, K., Rowe, R.E., Sommervile, G., Taylor, H.P.J., Threlfall, A.J., and Whittle, R

(1987) - *Handbook to British Standard BS8110:1985 – Structural use of concrete*, page 15, Palladian Publications, London.

Alireza Bagheri, Abbas Ajam, Hamed Zanganeh (2019) - Investigation of chloride ingress into concrete under very early age exposure conditions, https://doi.org/10.1016/j.conbuildmat.2019.07.225 (Accessed: August 2020)

Beushausen, H. (2014) - *Principles of Performance-based approach for concrete durability*, RILEM International workshop on performance-based specification and control of concrete durability, Zagreb

Bickley, J., Hooton, H.D., and Hover, K.C. (2006) - *Preparation of a Performance-based Specification for Cast-In-Place Concrete,* RMC Research Foundation

Bijen, J. (2003) - *Durability of engineering structures – Design, repair and maintenance*, page 55, Woodhouse Publishing

Bittner, A., Christelis, G., Dierkes, K., Lohe, C., Matengu, B., Ó Dochartaigh, B.É., Quinger, M., & Upton, K. (2016). *Africa Groundwater Atlas: Hydrogeology of Namibia*, http://earthwise.bgs.ac.uk/index.php/Hydrogeology_of_Namibia, Accessed February, 2019

Breugel, K, Polder, R., and Wegen, G, (2012) - *Guideline for service life design of structural concrete-A performance based approach with regards to chloride induced corrosion*, pages 153-164, Heron vol.57, 2012

Buchanan, A.H., Gibson, T., and Morris, H., (2018) - *Fifteen years of performance based design in New Zealand*, University of Canterbury, https://ir.canterbury.ac.nz/handle/10092/30 , Accessed 2018.

Byers, B., (1997) - *Environmental threats and opportunities in Namibia: A comprehensive assessment*, Research discussion paper No. 21, page 4, Department of Environmental Affairs

Calaghan B.G. (1991), *Atmospheric corrosion testing in southern Africa – results of 20 year national exposure programme,* page 49, CSIR, Scientia publishers, Republic of South Africa.

Camarini, G., DaSilva, M.G., Gomes, V., and Tanesi, J., (2010) - *From prescription to performance: International trends on concrete specifications & Brazilian perspective*, pages 420-431, IBRACON Vol. 3 No. 4

Caoa Y., Gehlenb, C., Angstc U., Wanga L., Wanga Z., Yaoa Y. (2018) - Critical chloride content in reinforced concrete — An updated review considering Chinese experience https://doi.org/10.1016/j.cemconres.2018.11.020 (Accessed August 2020)

Choo, B.S. and Newman, J. (2003) - *Advanced Concrete Technology-Constituent Materials*, pages 1/15-1/22; pages 4/3-4/35 & 5/1-5/34, Elsevier Ltd

CIRIA C614 (2010) - *The use of concrete in maritime engineering– a guide to good practice*, page 95, CIRIA

COLTO – *The standard specifications for road and bridge works for state road authorities*, 1998 edition, Department of Transport in South Africa, Republic of South Africa

Concrete Practice Guidance on the practical aspects of concreting, Concrete Society and British Cement Association, 2008

Constitution of the Republic of Namibia, Namibian Constitution First Amendment Act, 1998

Christelis G. and Struckmeier W. (2011) Department of Water Affairs, Ministry of Agriculture, Water and Rural Development, the Geological Survey of Namibia, Ministry of Mines and Energy, the Namibia Water Corporation-Groundwater in Namibia-an explanation to the Hydrogeological Map, Second edition, Windhoek, 2011.

Edvardsen, C., (2010) - *The Consultant's view on service life design*, page 249-264, 2nd International symposium on service life design for infrastructure, Delft

Edvardsen, C., Siemens, T., (1999) - *Duracrete: service life design for concrete structures-A basis for durability of other building materials and components?*, Proceedings of the 8th International Conference on Durability of Building Materials and Components, 8dbmc, Volume 2: Durability of Building Assemblies and Methods of Service Life Prediction, 1343-1356, Ottawa

Ekolu, S.O. (2018) - *Effects of superplasticisers on concrete durability indexes*, https://core.ac.uk/download/pdf/54201381.pdf, Accessed 2018.

Ekolu, S.O. and Murugan, S. (2012) - *"Durability Index Performance of High Strength Concretes Made Based on Different Standard Portland Cements,"* Advances in Materials Science and Engineering, vol. 2012, http://dx.doi.org/10.1155/2012/410909, Accessed 2018

Fib bulletin 34 (2006) – *Model Code for Sevice life design*, page 26-31, International Federation for Structural Concrete

Folic, R. and Grokovic, S. (2015) - *Some aspects of designing concrete structures based on durability*, 1st Scientific applied conference with international participation, page 267-276, University of Architecture, Civil engineering and Geodesy-Sofia, Bulgaria

Gjorv, O.E. (2013) - *Durability design and quality assurance of major concrete infrastructure*, pages 45-61, Advances in Concrete Construction Vol. 1 No.1, DOI: http://dx.doi.org/10.12989/csm.2013.1.1.045,

Google map 2019 - https://www.google.com/maps/place/Swakopmund+Sewage+Treatment+Plant/@-22.6630102,14.5347543,345m/data=!3m1!1e3!4m5!3m4!1s0x1c765bfe5aea2e95:0x355c8fa57d890b8!8m2!3d-22.6131129!4d14.5617737, Accessed 2019

Goudie, A., and Viles, H., (2015) - *Landscapes and Landforms of Namibia*; Springer Science+Business Media Dordrecht

Gun, J., v, d and Weert, F.v, (2012) - *Saline and brackish groundwater at shallow and intermediate depths: genesis and world-wide occurrence*, 39th International Association of Hydrogeoligist- International Congress, Niagara Falls, 2012

Haifeng Yuan, Patrick Dangla, Patrice Chatellier, Thierry Chaussadent (2015) - Degradation modeling of concrete submitted to biogenic acid attack, https://doi.org/10.1016/j.cemconres.2015.01.002 (Accessed August 2019)

Helland, S. (2016) - *Performance- based service life design in the 2021 version of the European concrete standards – Ambitions and challenges,* page 32-40, fib Symposium key note address, Cape Town

ISO 16204:2012, *Durability service life design of concrete structures*, The International Organisation of Standardisation, Geneva, 2012

Kudzai, K.F, Koeniger P, Turewicz V, Wang L, & Wanke H, (2016) *An Analysis of Precipitation Isotope Distributions across Namibia Using Historical Data*, PLoS ONE 11(5): e0154598. doi:10.1371/journal.pone.0154598

Lancaster, N. (2002). How dry was dry? Late Pleistocenepalaeoclimates in the Namib Desert. Quaternary Science Reviews, 21, 769–782. doi:10.1016/S0277-3791(01)00126-3 (Accessed August 2019)

Lamond, J.F, and Pielert, J.H. (2006) - *Significance of Tests and Properties of Concrete and Concrete-Making Materials*, STP 169D, page 238-250, ASTM International.

Lemey, L., Lobo, C., and, Obla, K., (2006) - *Performance based specification for concrete, Architectural Engineering Conference*, Omaha, 2006

Mehta, P., K., and Monteiro, P.J.M., (2006) - *Concrete microstructure, properties and materials,* pages 203-204; page 254; pages 282-284, The McGraw-Hill Company

Ng P.L., Kwan A.K.H. (2015) Improving Concrete Durability for Sewerage Applications. In: Tse P., Mathew J., Wong K., Lam R., Ko C. (eds) Engineering Asset Management - Systems, Professional Practices and Certification. Lecture Notes in Mechanical Engineering. Springer, Cham. https://doi.org/10.1007/978-3-319-09507-3_89 (Accessed August 2019)

Namibia Economist (2016), *Cement production upped to 1 million tons per year, Namibia Economist,* https://economist.com.na/21229/mining-energy/cement-production-upped-to-1-million-tons-per-year/, Accessed December 2016

NAMPORT – Namibia Ports Authority, *NAMPORT marine concrete specification*, Namibia

Ohorongo Cement (2019) - *Ohorongo Cement Indicators*, https://www.ohorongo-cement.com/cms_documents/types-of-ohorongo-cement-and-the-indicators-on-ohorongo-cement-c430d9db9b.pdf, Accessed August 2019.

Perkins, J.H. (1997) - *Repair, Protection and waterproofing of concrete structures*, page 48-50, 3rd Edition, E&FN Spon

Portland Cement Association (2002) –Types and Causes of Concrete Deterioration, PCA R&D Serial no. 2617, page 10, Portland Cement Association, Illinois

RA (2019), Namibia Roads Authority, Public notice, https://www.facebook.com/notes/roads-authority-namibia/closure-of-a-section-of-sam-nujoma-avenue-in-swakopmund/2259823847443416/, Accessed 2019

SANS 1200 (1986) – SANS 1200, *Standardised specification for civil engineering construction*, South African Bureau of Standards, South Africa, 1986

Santhanam, M. (2013) - *Durability and Performance Specification*, page 100-114, The Masterbuilder, 2013.

South African National Roads Agency Ltd (2019) - *Concrete Durability Specification Targets* (Civil Enginering Structures only) https://www.nra.co.za/content/Table_60001_Concrete_Durability.pdf, Accessed 2019

University of Cape Town (2017), Durability Index Testing Procedure Manual Ver 4.0, COMSIR, Cape Town, Feb. 2017

van Dyk, W.D. (2012) - Civil Engineer October 2012 Edition, *Swakopmund Municipality's new wastewater treatment plant*, SAICE, 2012.

www.ezilon.com (2017) – Political map of Namibia, https://www.ezilon.com/maps/africa/namibia-maps.html (Accessed: August 2020)

www.structurae.net (2019), https://structurae.net/structures/swakop-river-bridge, Accessed 2019

Yan-Chyuan Shiau, Chih-Ming Huang, Ming-Teh Wang, Jin-Yi Zeng (2008) - *Discussion of pot bearing for concrete bridge*, The 25th International Symposium on Automation and Robotics in construction, 26-29th June 2008, Lithuania

Yarong Song, Elaine Wightman, Jagadeesh kumar Kulandaivelu, Hao Bu, Zhiyao Wang, Zhiguo Yuan, Guangming Jiang (2020) - Rebar corrosion and its interaction with concrete degradation in reinforced concrete sewers, https://doi.org/10.1016/j.watres.2020.115961 (Accessed: August 2020)

Yarong Song, Elaine Wightman, Jagadeesh kumar Kulandaivelu, Hao Bu, Zhiyao Wang, Zhiguo Yuan, Guangming Jiang (2018) - Corrosion of reinforcing steel in concrete sewers, https://doi.org/10.1016/j.scitotenv.2018.08.362 (Accessed: August 2020)

Appendices

Appendix 1 -Research questionnaire

QUESTIONNAIRE

1. **CLIENTS**

a) Which part of the concrete construction industry is your firm/organisation in?

Public or Private

b) Which area of concrete works are you involved in?

Road structures, building structures, both

c) Which region of the country are your works in?

Name region or all over the country

d) If all over the country which region do you experience the most concrete deterioration issues?

Name region or none

If yes then provide brief detail of deterioration

e) If you are based in specific region have you experienced earlier than anticipated concrete deterioration in structures developed in your region?

Yes or No

If yes provide brief detail of deterioration

f) How do you determine the life of the concrete structure before you undertake maintenance?

Designer recommendation, Financial model, Building service life model, Ad hoc

g) Do you carry out any tests to confirm built structure will last in line with design life?

Yes or no

If yes indicate tests done

2. CONCRETE DESIGNERS AND CONSTRUCTION WORK SUPERVISORS

a) Which part of the industry are you in?

Private or public

b) Which concrete structures is your firm involved in?

Road structures, building structures, both

c) How do you determine the design life of concrete structures?

Client instruction, design codes guidance, building regulation guidance, service life design model, ad hoc

d) Do you undertake specific durability design for concrete structures?

Yes or No

If yes which durability model do you use?

If yes which design codes do you use for durability design

SANS, BS, Eurocode, American code, none of above

If none of above state code used

e) Have you had early concrete deterioration on your projects?

Yes or No

f) Were the deterioration restricted to a specific region?

Yes or no

If yes state the region

g) In case where you have experienced deterioration what were the common deterioration mechanisms you experienced?

State deterioration mechanism per area

h) How do you specify for durability?

Limiting cover, limiting water/cement ratio, limiting strength, limiting cement quantity, reinforcement detailing, durability indices (select as many as possible)

If option not included add

i) How do you check for concrete quality to ensure durability?

Reinforcement inspection, slump test, strength test, cover measurement, durability indices (select as many as possible)

If option not included add

j) Do you verify durability quality of AS-Built concrete?

Yes or No

If yes how do you do it? ...visual inspection, cover measurement, coring, durability indices

k) Do you put in testing requirements in recommending a maintenance monitoring plan for concrete

Yes or no

If yes what tests are included?....visual inspection, cover measurement, coring, durability indices

l) Would you consider undertaking CPD for your design team on performance based durability design and specification?

Yes or No

If no why?

m) Have you requested for specific durability related tests on your projects?

Yes or no

If yes which tests?

Were the test done by local or international laboratory?

3. LABORATORY

a) Are you a regional or national materials laboratory service?

National or regional

If regional ...name region

b) How long have you been offering concrete testing service?

c) Which areas of concrete testing do you provide service?

Material testing, concrete mix design, fresh concrete testing, in-situ concrete testing (choose option)

d) Do you provide mix design service

Yes or No

If you offer concrete mix design service which specification do you follow?

Engineer's drawing specification, standard specification, develop your own (choose option)

If you follow standard specification specify which one

e) Are you familiar with the South African durability tests?

Yes or no

If YES go to (e)

If NO go to (f)

f) Do you provide any concrete testing for durability indices?

Yes or No

If yes which one

g) Would you be interested in doing CPD for concrete durability

h) Which hardened concrete testing do you have capacity to provide?

Coring, chloride profiling, durability indices, cover measurement

4. CONCRETE PRODUCERS

a) Do you produce concrete in one place or you have multiple batching places?

One or multiple

If multiple are they all in the same region?

If in multiple regions which ones?

b) Do you use an external laboratory for concrete tests?

Yes or No

c) Do you do your own concrete mix design or you use external expertise?

Internal or external

d) Which specifications do you follow in concrete mix design?

SANS, BS, Eurocode, Project specifications, drawing specification others

If others specify

e) What measure of performance do you normally use?

Strength, slump, durability indices, workability or others

If others specify

f) Is the quality of the concrete you produce given any relationship to the life of the structure to be built?

Yes or No

If yes how

g) Do you undertake concrete placement?

Yes or No

If no then do not proceed

h) What tests do you do on site before you place concrete for acceptance?

Slump, cube sampling

i) What works do you do on site you consider vital for concrete durability

Transportation, vibration, curing, shuttering, others (choose as many)

If others specify

j) Have you worked on projects where you have been asked to test cast concrete to confirm durability of concrete?

Yes or No

If yes how many times?

k) Are you familiar with performance based concrete construction specification?

Yes or No

If No, would you consider undertaking CPD in this area?

www.ingramcontent.com/pod-product-compliance
Lightning Source LLC
LaVergne TN
LVHW091034150826
845672LV00006BA/1808

* 9 7 8 3 3 8 4 2 2 4 7 7 4 *